NEW DIMENSIONS IN MANUFACTURING

By
Dr. Bert P. Erdel

Hanser Gardner Publications
Cincinnati

Erdel, Bert P.
  New dimensions in manufacturing / by Bert P. Erdel.
     p.   cm.
  Includes bibliographical references and index.
  ISBN 1-56990-245-3 (hardcover)
  1. Manufacturing processes. 2. Machining. 3. Production management.
4. Concurrent engineering. I. Title.
TS183.E73 1998
670.42--dc21   98-26023
                                                                CIP

While the advice and information in *New Dimensions in Manufacturing* is believed to be true, accurate, and reliable, neither the author or the publisher can accept any legal responsibility for any errors, omissions, or damages that may arise out of the use of this information. The author and publisher make no warranty of any kind, expressed or implied, with regard to the material contained in this work.

Hanser Gardner Publications
6915 Valley Avenue
Cincinnati, OH 45244-3029

Copyright © 1998 by Hanser Gardner Publications. All rights reserved. No part of this book, or parts thereof, may be reproduced, stored in a retrieval system, or transmitted in any form or by any means without the express written consent of the publisher.

1 2 3 4 5 6   01 00 99 98

TABLE OF CONTENTS

Preface

Acknowledgements

Chapter 1:
Productive and Cost-Effective High Speed/
High Velocity Machining 1
    The Machine 3
        Machine Features — Bearings 3
        Machine Features — Motors 4
        Open Architecture 5
        Heat, Dust, and Chip Control 7
        Fast Tool Changes and Rapid Pallet Exchange 7
        Mobility and Ease of Maintenance 7
        Part Fixturing 8
        Manufacturing Cells 8
    The Tooling Interface 8
        Constant Uniform Clamping, and Static Dynamic Stiffness . 10
        Transmission of High Torque 11
        Repeatability Accuracy, and Positional Accuracy 11
        Short Overall Length 14
        Length, Weight, and Diameter 14
        Allowable Unbalance 15
        Standardization 15
        Manual and Automatic Tool Change 17
        Speed Limits 17
        Adaptability 17
    Cutting Tools 20
        Workpiece Material/Cutting Operations 21
        Cast Iron and Steel 21
        Superalloys 22
        Aluminum and Magnesium 24
        Composites 25
        Cutting Material 26
        Special Design Features 31
        Applied Tool Management 34
        The Level of Machine Technology 35
        Optimum Cutting Speed 37
    Conclusion 41

Chapter 2:
Dry-Machining and Near-Dry-Machining 43
- Dry-Machining 46
  - Cutting Material 46
  - Tool Design and Machining Data 49
  - Workpiece 50
  - Machine Tool 52
  - Applied Technology 52
  - Turning 53
  - Reaming/Fineboring 56
  - Drilling 59
  - Milling 62
  - "Dry and High" 63
- (Near)-Dry-Machining 63
  - Applied Technology 67
  - Drilling 67
  - Reaming 67
  - How Does MVL Fare in Finish-Machining Aluminum Bores? 70
- Summary 72

Chapter 3:
Precision One-Pass Machining 75
- From Traditional to Advanced Machining 76
- Advanced Machining Technology 78
  - Cutting Material, and Workpiece Material 78
  - Optimized Cutter Geometry 81
  - Well-Defined and Balanced Tool Body 82
  - Accurate Toolholding, and Tooling Interface 85
  - Process Monitoring 85
- Fine-Tuning of the Tool Assembly 87
- Substituting Processes 89
- Applications in Practice 92
  - Hardturning 92
  - Circular Milling/Circular Interpolation 95
  - Fineboring 97
  - Advanced Generating 99

Chapter 4:
Global Concurrent Manufacturing 103
   Time, Quality, and Cost 103
   The Relevance of Concurrent Engineering —
   Its Purpose and Principles 104
      Product Life Cycle 107
      Opportunity 108
      Strategy 108
      Innovation 108
      Design 108
      Prototyping 108
      Manufacturing 109
      Continuous Improvement 109
   Applied Concurrent Engineering: Leading the Way to
   Concurrent Manufacturing — A Case Study 110
      Saturn — A New Approach to Manufacturing 110
      Saturn's Applied Concurrent Engineering 111
   Concurrent Manufacturing 114
      The Concurrent Manufacturing Team 115
      Core Principles of Concurrent Manufacturing 116
   Global Manufacturing 122
      Strategic Considerations 123
      Managing Global Manufacturing 124
      World-Class Manufacturing 126

Chapter 5:
No-Nonsense Manufacturing 129
   Introduction 129
   Supply Chain Effectiveness — The Competitive Edge 130
   Criteria for Choosing a Supplier 131
      Supplier Types and the Advanced Flow of Supplies 131
      Challenges and Opportunities for Suppliers 133
      The Manufacturer — Self Induced Risks 137
      Making it Work — The Advanced Supply Chain 139
      Supplier Integration 141
   Striving for Simplicity and Conquering Complexity 143
      Design, Process, Product 143
      Automation and Simplicity 146
      Part- and Product-Oriented Manufacturing 148
      Managing Complexity 150

  Measures of Advanced Management 155
   Lean and Team 155
   Changes and Adjustments 160
   Revisiting "Re-engineering" 160
   The Process of Change 163
   Adjustments 169
  Agility — The Journey of Continued Competitiveness 171
   Drivers of Change and Competitiveness 171
   Key Attributes of the Agile Manufacturing Company 173
   Characteristics of Agility 173
   Customer Information/Communication 174
   Measuring the Progress of Agility 177
   New Frontiers — New Challenges 177

Conclusion 179
References 181
Index 183

PREFACE

The business world is changing at an ever increasing, breathtaking pace. Many companies are having difficulties adjusting and keeping up, and are being left behind. In corporate boardrooms, there is worry of being overwhelmed by competition or falling victim to the marketplace.

I decided to write this book to raise the awareness of manufacturing companies — to spread the word that there are advanced processes, techniques, strategies, and principles that hold the key for staying ahead in a world of constant technological advancement and perpetual adjustments.

We need to realize that power has shifted from the manufacturer to the customer who now dictates what, when, and how to produce products. The old paradigm of rigid production runs, traditional organizational hierarchies, and conventional manufacturing processes must be revisited. Exchanging outdated, archaic, manufacturing techniques and technologies for advanced technologies such as high velocity machining, dry-machining, and one-pass machining can yield dramatic productivity increases and allow companies to rapidly and economically respond to customer demands.

Technology alone, however, cannot secure corporate growth and well being. The core competency of any company is its knowledge, and the level of knowledge is determined by the most precious resource: people. To better utilize the human mind and its enormous potential within group settings, team building united with open communications and the free flow of information are important ingredients for success. Multidisciplinary teams can be formed to improve products and processes, brainstorm for new technology, benchmarking, cost cutting, quality assurance, and continuous improvement.

The ever-growing complexity of processes and products does not permit any single manufacturer to go it alone. So much knowledge and expertise cannot be gathered and groomed in isolation, but must be obtained in concert with other expert companies. This interdependency and the shrinkage of the world through communication networks have produced a global manufacturing and marketing arena of gigantic proportions. Only the leanest, fittest and most agile companies will be able to survive the onslaught of fast-changing technology and rapid market-changes.

It is clear that the pace of today's business will speed up before it gets any slower. We must understand that what we do well today may become insufficient and outdated tomorrow. To stay on top is to anticipate gyrations, shifts, and changes, and adjust to them with knowledge and know-how to improve upon them by utilizing the best technology and human resource techniques available.

Bert P. Erdel

ACKNOWLEDGEMENTS

New Dimensions in Manufacturing is the summary of my direct involvement in numerous manufacturing activities undertaken by industry, academia, and government. A great number of distinguished, hardworking, and dedicated professionals have inspired me to write this book. To all of them I am deeply thankful.

My special thanks go to the many customers who have given me the benefit not only of their encouragement, but also their invaluable insights and experience gained from day-to-day operations.

I would also like to acknowledge the support of my staff, particularly David Itterly for the superb artwork, and Erika Scheffel for her patience and support throughout the development of the book.

Bert P. Erdel

DEDICATION

To My Beloved Children
Gillian, Wolfgang, Christoph

Chapter 1

PRODUCTIVE AND COST-EFFECTIVE HIGH SPEED/HIGH VELOCITY MACHINING

Today's manufacturers must restructure and retool for a new era of fierce competition. Participants in the emerging global marketplace will come out ahead only if they are able to respond rapidly to market shifts and changing customer demands. Manufacturers must become more flexible and agile in order to accommodate new product mixes, varying production quantities, and expanded part families. Quality's role will not diminish — it will continue to be as important as cost-effectiveness and productivity.

Processes that can sustain high degrees of productivity will dominate if they offer the potential to enhance quality while remaining cost-effective. It is here that the complex technology of High Speed/High Velocity (HS/HV) machining is extremely promising. By synthesizing a myriad of technologies, HS/HV machining reduces cycle times, increases throughput, minimizes non-machining time, and reduces main machining time, and it *can* yield high precision with a favorable cost/performance ratio.

High speed, as defined by the rotational spindle speed, today spans a range of 15,000 to 25,000 rpm for typical machining operations. However, since the tool translates rotational (idle) spindle speed into practical use, the term "high cutting speed" is more appropriate. High cutting speed ranges between 3,000 and 5,000 m/min in nonferrous applications (most commonly when machining aluminum parts), but the process has also been applied to steel and cast iron when cutting at speeds of up to 1,000 to 1,200 m/min. In any event, the influence of high cutting speeds within the machining envelope is rather dramatic.

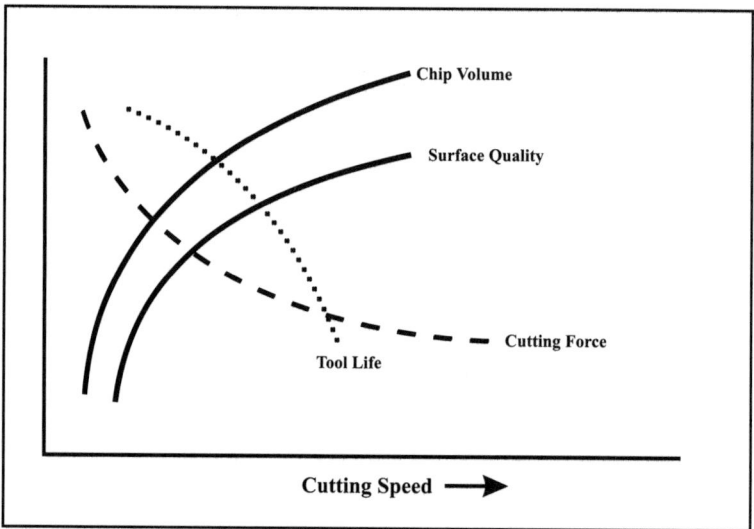

Figure 1.1 • Machining Characteristics

While chip volume and surface quality increase with higher cutting speeds, cutting tool life is shortened and cutting forces decrease, as shown in Figure 1.1. A smaller amount of the heat generated at the cutting point is transferred into the workpiece and cutting material than is normal with traditional machining because much of the heat is removed with the chips. The lower tangential loads on the cutting tool simplify machine and part fixture design, and allow for easy machining of thin-walled workpieces. Better surface finishes are, as a rule, a welcome bonus, as are the higher chip-load demands, well-defined tool chip galleys, and reliable transport of the chips out of the machining area. Advanced cutting tool material can prolong tool life to acceptable levels, and hard, heat-resistant cutting material can also keep reductions of tool life at acceptable levels.

To reap the benefits of this evolutionary machining concept, it is important to synthesize the technology triumvirate comprised of the machine, tooling interface, and cutting tools. All of the individual components that make up these three groups must be well designed and engineered to their respective highest levels and, in unison, must represent state-of-the-art design and workmanship. The most relevant criteria to consider are shown in Figure 1.2.

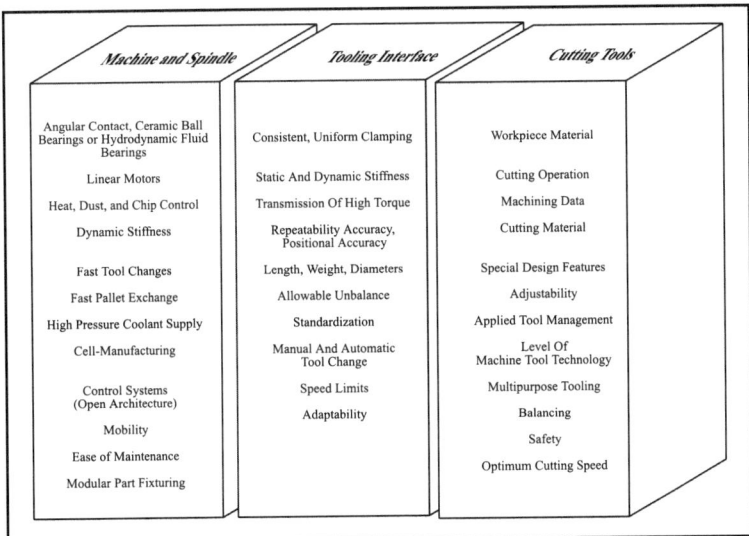

Figure 1.2 • Criteria of High Speed/High Velocity Machining

THE MACHINE

This section describes the critical features of a HS/HV machine. Figure 1.3 shows the module design of a machine of this type.

MACHINE FEATURES — BEARINGS

Angular contact bearings are popular — especially hybrid bearings (steel cage and ceramic balls) — because they provide minimal friction and wear as well as high load capacity of both axial and radial forces. Hydrostatic/hydrodynamic fluid bearings are excellent alternatives, especially for higher speeds. In addition, they offer excellent damping characteristics and low vibrations combined with low runout and virtually unlimited lifespan.

4 • Chapter 1

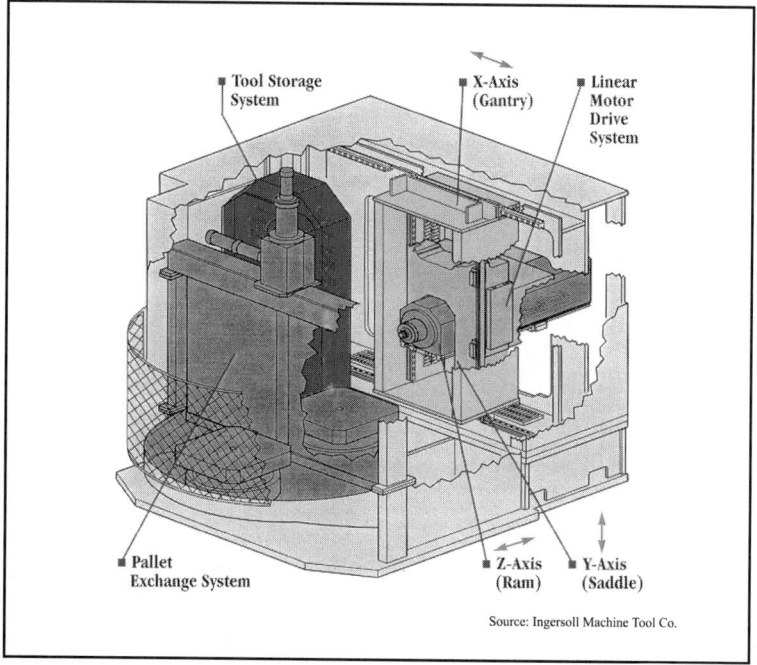

Figure 1.3 • Machine Module

Machine Features — Motors

Whenever high cutting speeds are to be applied, the machining capability should include fast axes movements and acceleration. These abilities take high speed machining closer to high velocity machining. Linear motors, driven in place of servo motor and ball screw, produce rapid traverse rates of 4,000 to 4,500 ipm, and acceleration and deceleration rates of up to 2.5 g's (See Figure 1.4). By comparison, a ball screw design delivers traverse rates of about 2,000 ipm max with 1.0 g deceleration, as well as accuracies of 20-25 mm. Ball screws perform adequately for high speed machining, but for the upper end of high velocity machining, only linear motors can provide the performance capabilities that are required. The fact that conventional machining centers can reach acceleration rates of about 1.0 g is due to advanced control systems that provide high speed processing of new algorithms as well as multiple and dedicated digital signal processing. Combined with these features, they possess new ball screw designs that allow for weight reduction (hollow shafts) and double ball screw architecture for accuracy and heat dissipation.

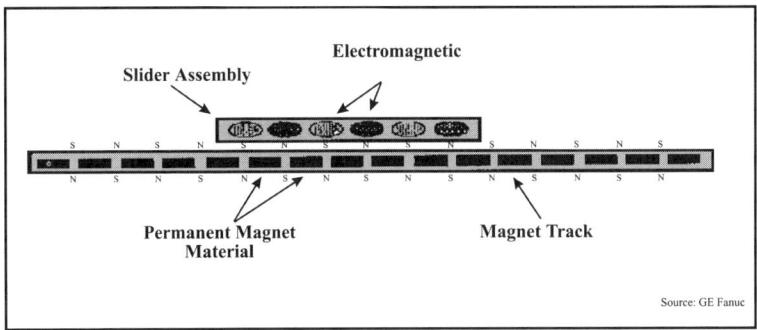

Figure 1.4 • Brushless Linear Motor Design

Still, the linear motor has an array of features that are ideally suited for high speed machining. In conjunction with digital control systems, the linear motor can achieve high speeds and excellent positioning accuracy, and is not handicapped by backlash, inertia, and the lack of elasticity that can limit static and dynamic stiffness and reduce the acceleration rate of ball screw arrangements. The contactless transmission of forces in linear motors, by contrast, reduces wear and allows virtually unlimited axis travel. Another important plus of the linear motor is that no machining takes place during acceleration.

The disadvantage of the linear motor is its low degree of efficiency and the resultant loss in output that leads to excessive heat being generated and stored within its structure. This, in addition to its tendency to attract dust and chips (due to its magnetic field), has led researchers to find ways to seal the magnet tracks, improve feedback devices, and to employ digital amplifiers to reduce heating. Still, linear motors are undoubtedly the wave of the future because they offer an important contribution to high velocity machining.

Open Architecture

All machine tools built for HS/HV machining have, of course, been designed for flexible automation and agile manufacturing. Control systems must be fast, accurate, and reliable. The emerging areas of ever more powerful programmable computers, client/server computing de-

6 • Chapter 1

sign, and open interfaces or open architecture, as well as standards like OLE and CORBA, will propel manufacturing further. HS/HV machining is rapidly making inroads into production, and the new technology is being mixed with standard, low- to medium-speed machining centers. Retrofitting existing machine tools so that they could all communicate with each other would be extremely costly and technically difficult to do. For this reason, the emerging high speed machine tool technology provides open interfaces and open controllers.

Standardized open architecture is needed for all machine tools to allow for integration of control and computer systems. It should be possible to take a newly acquired HS/HV machining system and "open" it up to other machines of different manufacture, and to fit it with different controls to maximize its flexibility and agility.

While the hardware of a machine tool might be customized for a specific workpiece and a specific industry, the software should be generic to facilitate communications. In order to attain flexibility, fast machine tools need to be designed independently of proprietary control hardware and software. Equipped with standardized hardware and basic software packages, end users can then customize the controllers themselves according to their needs on the production floor. The goal of openness for manufacturing is to allow the entire organization to interlink through open architecture.

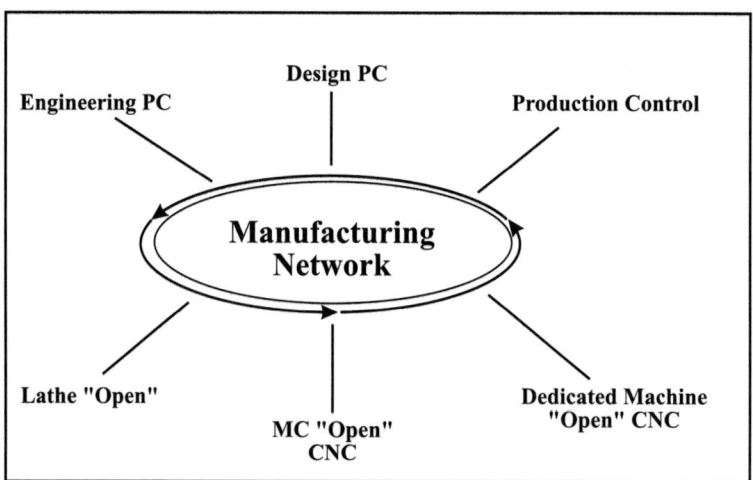

Figure 1.5 • True Open Architecture

Heat, Dust, and Chip Control

Machining at high speeds and with high velocity generates excessive heat within the area of the rotating and moving parts. Proper heat dissipation through cold airstreams and water jackets is as much a necessity as the need to prevent dust, chip particles, and mist from entering sensitive parts of the machine (electronics, drive systems, valves, etc.). The machine section that houses the moving and rotating parts should be slightly pressurized to contain harmful particles within the machining compartment enclosure. Since chip volume increases with cutting speed, a slant bed design as well as high coolant pressure will aid in removing chips and debris from the immediate cutting area.

Fast Tool Changes and Rapid Pallet Exchange

To reap the benefits of high speed and acceleration rates, tool changes should not take longer than a few seconds — i.e., the cut to cut time should not exceed 3 to 4 seconds — and the pallet exchange should take place in no more than double the time of the tool change (within 8 seconds).

Mobility and Ease of Maintenance

All moving parts and modules should be made of lightweight materials (aluminum, magnesium, composites, or titanium), and their geometric design should be optimized through finite element programs. The reduction of weight and mass results in a reduction of the energy needed to move those parts while accommodating the need for mobility in the machine. A design such as a foundation independent three-point base will reduce installation time and foundation costs, while simplifying the movement of machining systems on the production floor, and will help absorb vibrations due to the more stable base.

Given the complexity of the machines and the amount of capital expenditure required, ease of maintenance is an important consideration. Using modular designs will allow for fast interchange of parts and components. When a spindle malfunctions, it is essential that a complete replacement can be installed in a timely fashion and with relative ease, because production cannot afford to experience costly downtime. In order to avoid unnecessary power and current variations, it is advisable to equip each machine with an individual transformer to level or filter the current. For the "plug and play" principle — i.e., bringing machines from one part of the plant to another — mobility and maintenance play a crucial part.

Part Fixturing

HS/HV machining typically takes place in the context of lighter cuts, mostly in nonferrous metals. Since the tangential force during cutting is smaller than with conventional machining, part fixturing should be designed to be as light as possible and to have low clamping pressure. However, machines have to be designed to accommodate parts with a variety of material composition and stock removal parameters. So the issue is not just how light the design can be, but if its flexibility and modularity is achieved at the expense of rigidity or reliability.

Manufacturing Cells

Especially in the manufacture of automotive components, HS/HV machines are grouped to form manufacturing cells of 4, 8, or 12 machines for high volume part families. Such arrangements replace conventional transfer lines for almost equal part throughput, but in a flexible, agile mode. The complete integration of hardware and software markedly broadens the scope of such systems, especially since there is a major shift toward just-in-time delivery and make-to-order concepts.

HS/HV allows manufacturers to work with speed and flexibility while remaining competitive. Beyond that, applied data collection and new organizational functions — including Enterprise Resource Planning (ERP) and Advanced Planning and Scheduling (APS) — induce accelerated response times and minimize the changeover time from product to product.

THE TOOLING INTERFACE

Against the backdrop of high speed, it becomes apparent that conventional tool shanks and tool holders will be unable to provide reliable and accurate machining. This is particularly true for the extremely popular ISO-taper that has dominated the tooling interface scene for decades. Its main disadvantages are its mass, lack of face or shoulder contact, tendency to get sucked up into the machine spindle due to the expansion of the spindle nose at high speeds, and accurate tool gripping force cannot be guaranteed. (See Figure 1.6).

These shortcomings of the ISO-taper have led to the development of the Hollow Taper Shank "HSK" — the most advanced mechanical tooling interface to date. In fact, it has the potential to revolutionize chip making. The key to HSK is that it locates simultaneously on both a slight

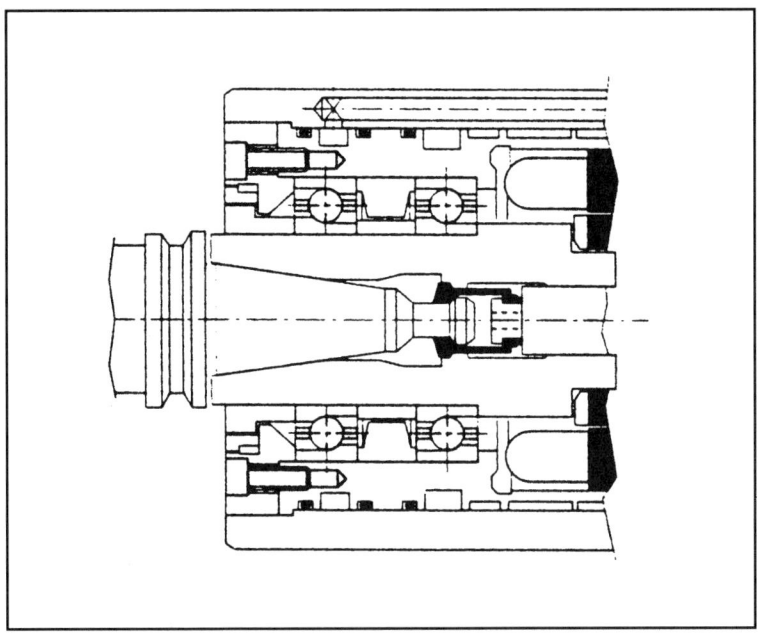

Figure 1.6 • V-Flange Clamping System

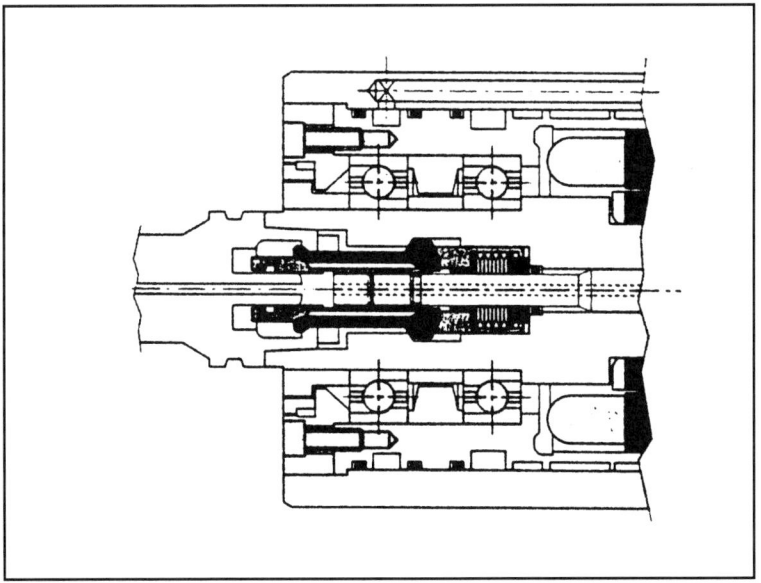

Figure 1.7 • HSK Clamping System

socket taper and a flat shoulder face, which is perpendicular to the spindle axis (see Figure 1.7). When the shank is pulled into the receiver (machine spindle or adapter), the design assures consistently uniform mating of machine and tool, as well as faster, more reliable tool changes.

Another feature, particularly important to high speed, is that the hollow HSK body design allows for slight expansion as speeds increase, causing the HSK system to "lean" against the inner wall of the spindle socket or the inner diameter of the corresponding clamping chuck, assuring dynamic stiffness. As the speed decreases, the interface contracts simultaneously with the spindle or holder. As we discuss the characteristics of the criteria for reliable HS/HV machining according to Figure 1.2, we will have a better understanding of why the HSK design is today's interface of choice.

Constant Uniform Clamping, and Static and Dynamic Stiffness

With the HSK system, the clamping forces are in the axial and vertical direction along the taper as well as through the face contact (Figure 1.8). This ensures consistent and uniform clamping. The flange/face contact ensures consistent gauge lengths from tool to tool, which means

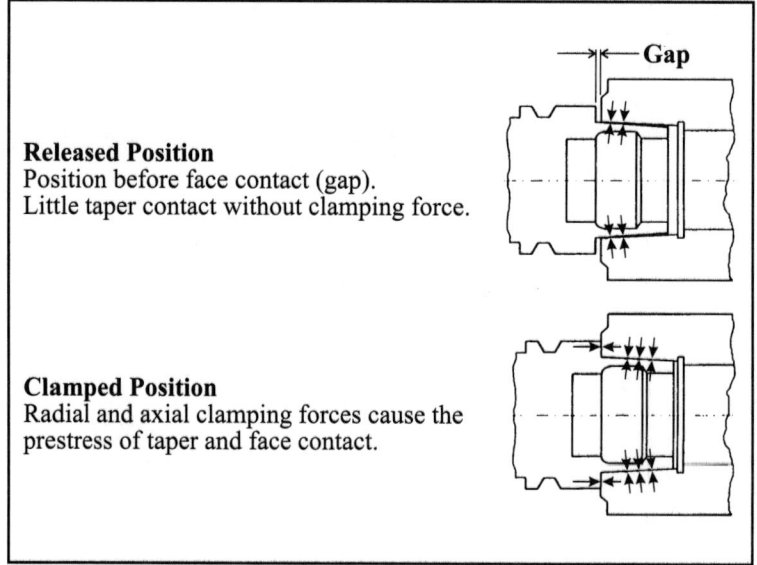

Figure 1.8 • HSK Distribution of Force

that pre-settings have to be programmed only once. The simultaneous mating of taper and shoulder face also gives the HSK its stiffness. Because the toolholder is a few micrometers shorter than the spindle socket, their mating results in a mechanical preload between face and spindle. This full contact of the mating surfaces makes the HSK highly resistant to radial deflections, hence its high stiffness. This high degree of static and dynamic stiffness translates into predictable machining results, even when machining parameters change, e.g., light cuts versus heavy cuts, or when machining with interruptions. In other words, it stabilizes the machining process, which results in minimal machining variations and eliminates possible chatter at the cutting edge.

Transmission of High Torque

When machining at high speed or with heavy cuts, the torque from the machine to the cutting tool must be securely transmitted. This can only be achieved through form and pressure clamping. In the event of varying loads during the cutting process that can result in erratic clamping forces, the interface could lose its defined position because of plastic or elastic deformation, and finish-machining within certain tolerances would be impossible to guarantee.

The transition from power clamping to form clamping takes place as the torque of the machine increases, and it is usually induced by the relative motions of spindle and interface. With higher pressure clamping, form clamping is not as necessary, and the fit of the connection becomes more accurate.

The hollow taper shank HSK transmits torque through its taper and face as well as its inner shoulder (pressure clamping). When this transmittable torque exceeds its limit, drive keys ensure form clamping to guarantee secure torque transmittal. A total of 10% of the clamping forces are transmitted through the taper, while 90% are transmitted through face contact. See Figure 1.9.

Repeatability Accuracy and Positional Accuracy

The axial and radial accuracy of the cutting tool is a function of spindle and interface accuracy. Axial and radial out-of-roundness conditions lead to unacceptable geometric tolerances, especially when tight SPC inspections demand that high tolerances have to be kept consistent over extended "sigma" runs, and "T.I.R." (Total Indicator Readouts) have to be kept to a minimum.

12 • Chapter 1

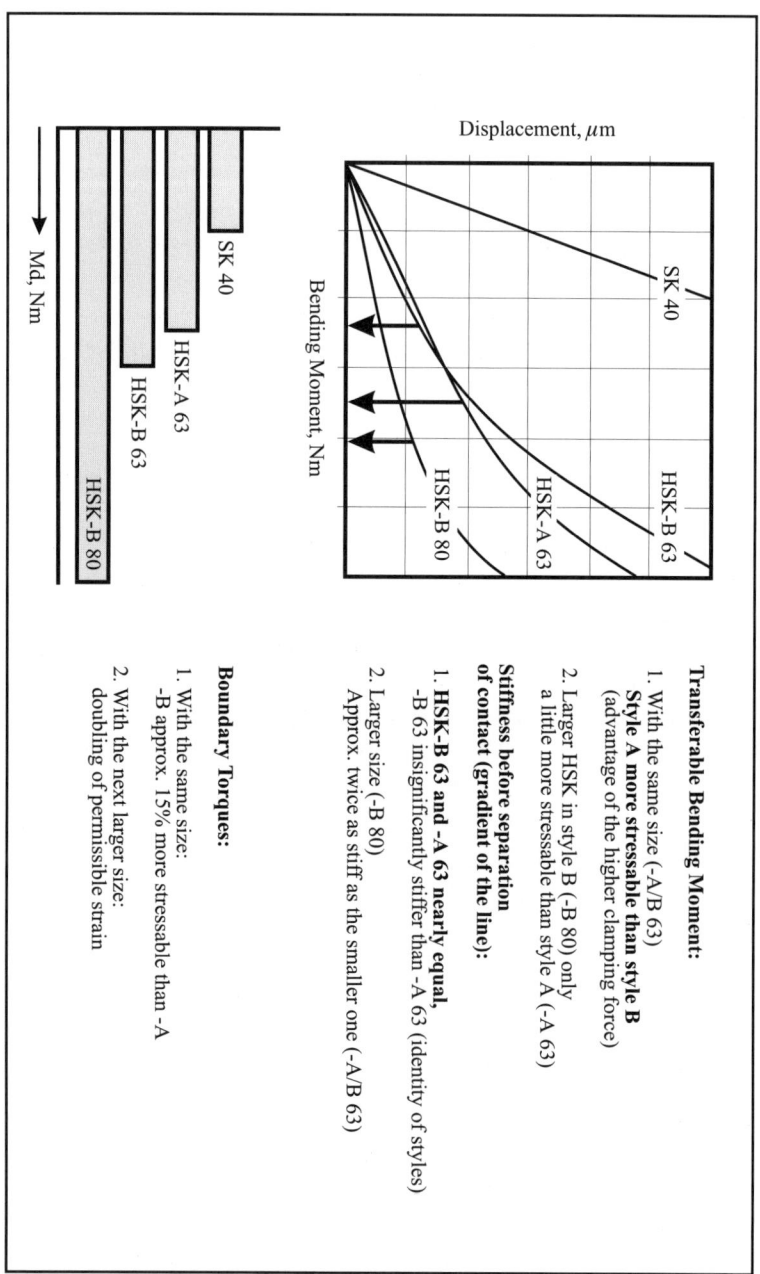

Figure 1.9 • HSK Bending Moment and Torque

Chapter 1 • 13

Since flexible manufacturing is characterized by frequent, repeated tool changes, machining has to be done with a high degree of repeatability accuracy. Both positional and repeatability accuracy are a direct function of machine spindle and tooling interface. Test runs have empirically determined the superiority of the HSK design, as shown in Figure 1.10.

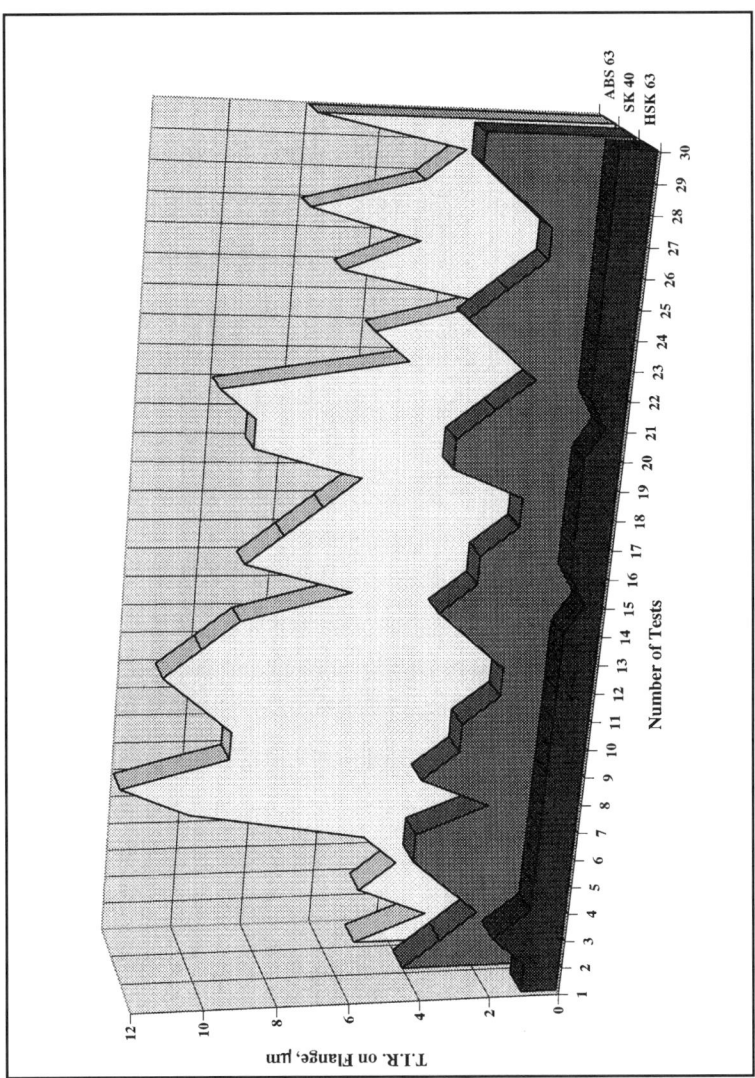

Figure 1.10 • Repeatability Accuracies

Short Overall Length

Figure 1.11 shows the interrelationship of T.I.R. (runout) to tool life and part tolerance. The smaller the T.I.R., the closer the part tolerance and the longer the tool life.

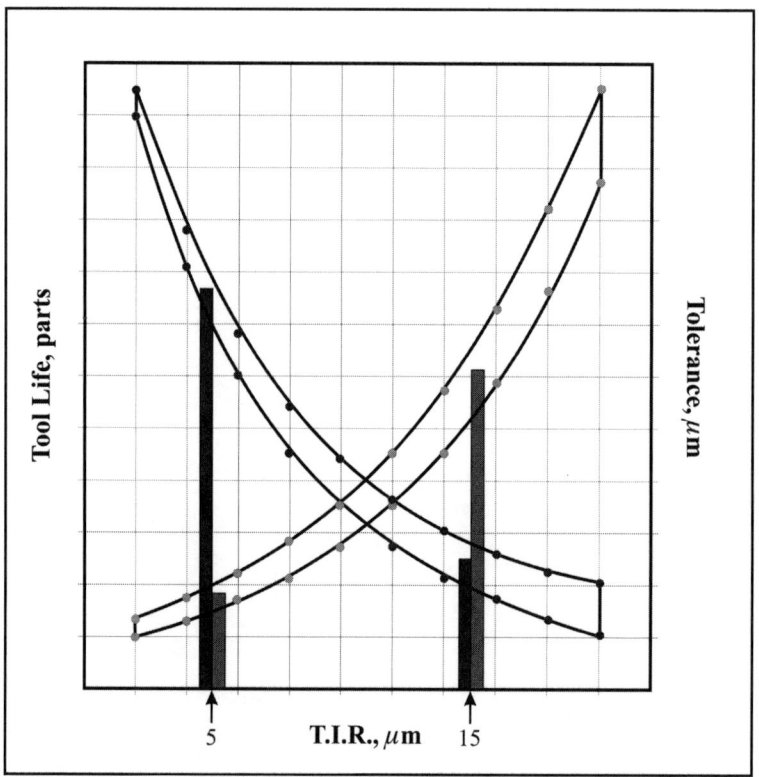

Figure 1.11 • Dependence on T.I.R. for Tool Life and Part Tolerance

Length, Weight, and Diameter

The short dimensional length of the interface allows for a short flow of force and a smaller bending moment. Therefore, the tool and holder combination is shorter, the load on the machine is decreased, and deflection during cutting is reduced.

When machining at high rotational speeds, centrifugal force becomes an important factor. Keeping the mass and weight to a minimum, and keeping the diameter as small as possible, are important design considerations. The shorter length of HSK also reduces tool change time, since the tool change arm doesn't have to pull the HSK out as far to clear the spindle as it would with a conventional taper shank.

The gauge line diameter of HSK is smaller than the CAT-V-flange, while the actual flange diameter of HSK is bigger. This generates a higher permissible bending moment. The smaller gauge line diameter allows for a smaller tool grip area, and the larger flange diameter allows for a higher radial load and higher feed rates. This is especially advantageous for milling operations.

Allowable Unbalance

As spindle speeds move above 10,000 rpm (which might be considered the low end of high speed machining), balance becomes a major issue. The relationship between the centrifugal force and speed is squared, i.e., doubling the speed quadruples the centrifugal force. An increase in clamping force accompanies increased speed (centrifugal force) and tightens the connection to the tool to enhance the system's stiffness, which is an important element of a balanced system. Machine tool, cutting tool, and tooling interface have to be balanced individually. As a system, however, it can show unbalances beyond allowable values if an excessive out-of-round condition prevails due to axial or radial inaccuracies. The high speed ranges are:
- low end = 10,000 - 15,000 rpm
- medium = 15,000 - 20,000 rpm
- high end = 20,000 rpm and up.

A rotationally symmetric design is essential, as is the need to manufacture the interface, through grinding, to the smoothest possible surface finish.

The ISO-standard 1940 determines balance specifications through grades. For most high-speed applications, the G2.5 grade is sufficient. For the high end, depending on the machining operation and system, a G1 grade might be specified. The maximum allowable imbalance as per the respective grade can be taken from Figure 1.12. A well-balanced interface is also an indicator of the system's safety, which is always an issue with high rotational speeds.

Standardization

The development of the hollow taper shank HSK not only led to the replacement of the conventional taper shank for high-speed machining,

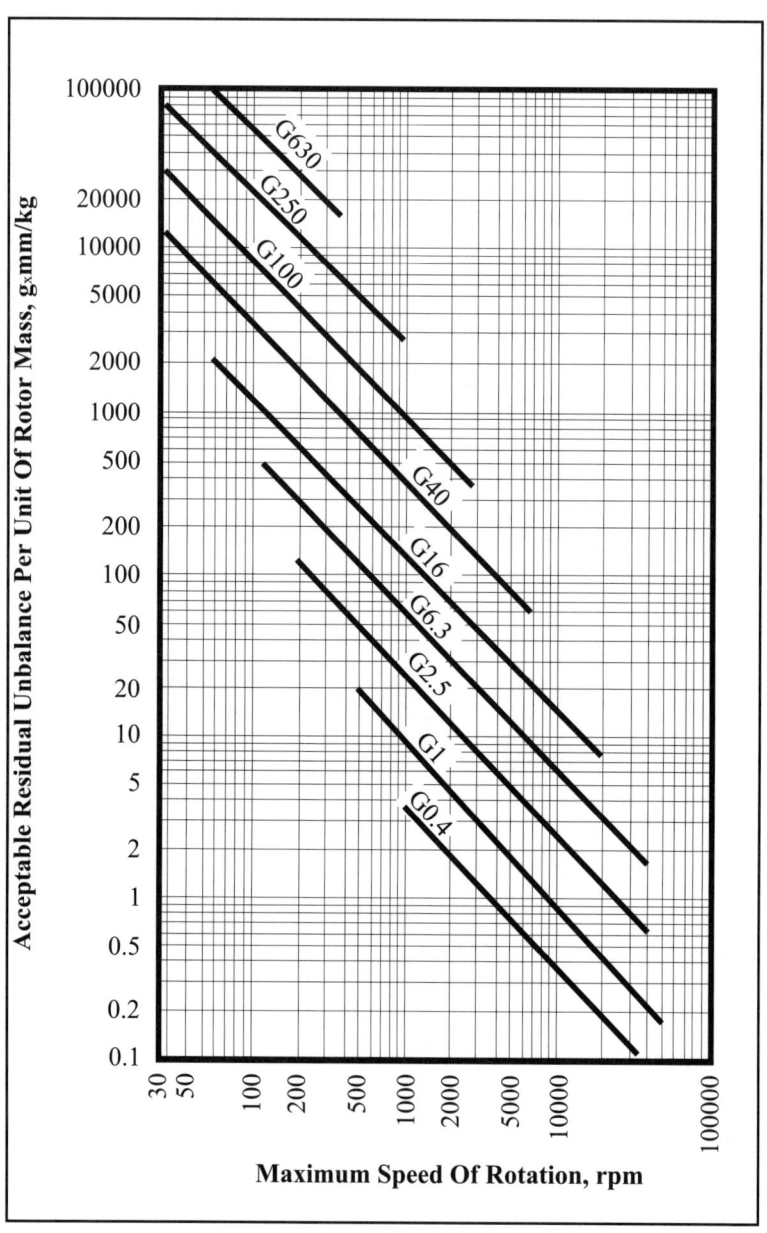

Figure 1.12 • Quality Grades

it also opened up a new avenue for standardization. HSK hollow shanks are defined as follows:
- HSK-A — automatic tool change
- HSK-B — automatic tool change (large shoulder diameter)
- HSK-C — manual tool change
- HSK-D — manual tool change (large shoulder diameter)
- HSK-E — automatic tool change for extreme high spindle speed (w/o drive keys)
- HSK-F — automatic tool change for extreme high spindle speed (w/o drive keys, large shoulder diameter).

HSK thus became a standardized, generic tooling interface.

Manual and Automatic Tool Change

A standardized tooling interface of the same basic design is needed for both manual and automatic tool change. HSK-A (except for the gripper groove for automatic tool change), operates on the same principle as HSK-C and, as a result, can also be used for manual tool changes. The clamping principle for automatic tool changer HSK-A is illustrated in Figure 1.13, and the clamping principle for the manual change HSK-C is shown in Figure 1.14.

Speed Limits

Determining the maximum allowable speed at which the tooling interface can reliably and safely operate is not just a matter of interface design and manufacture: the cutting tool being used and the spindle design are also part of the equation.

As for HSK, preliminary conservative diagrams showing the applicable speed limits take into consideration typical spindle diameters and designs (bearings, lubrication). Figure 1.15 shows the speeds that HSK can reliably achieve. The failure characteristic for HSK itself would be the loss of contact at the large taper shank diameter due to centrifugal force.

Adaptability

Any successful cutting tool interface has to be adaptable to other interfaces. This is especially critical with a newly created interface such as HSK. Its design fulfills the criteria for high speed machining, and it is easily adaptable to other reputable tool connections in wide use — an important criteria, indeed, since the vast majority of today's interfaces

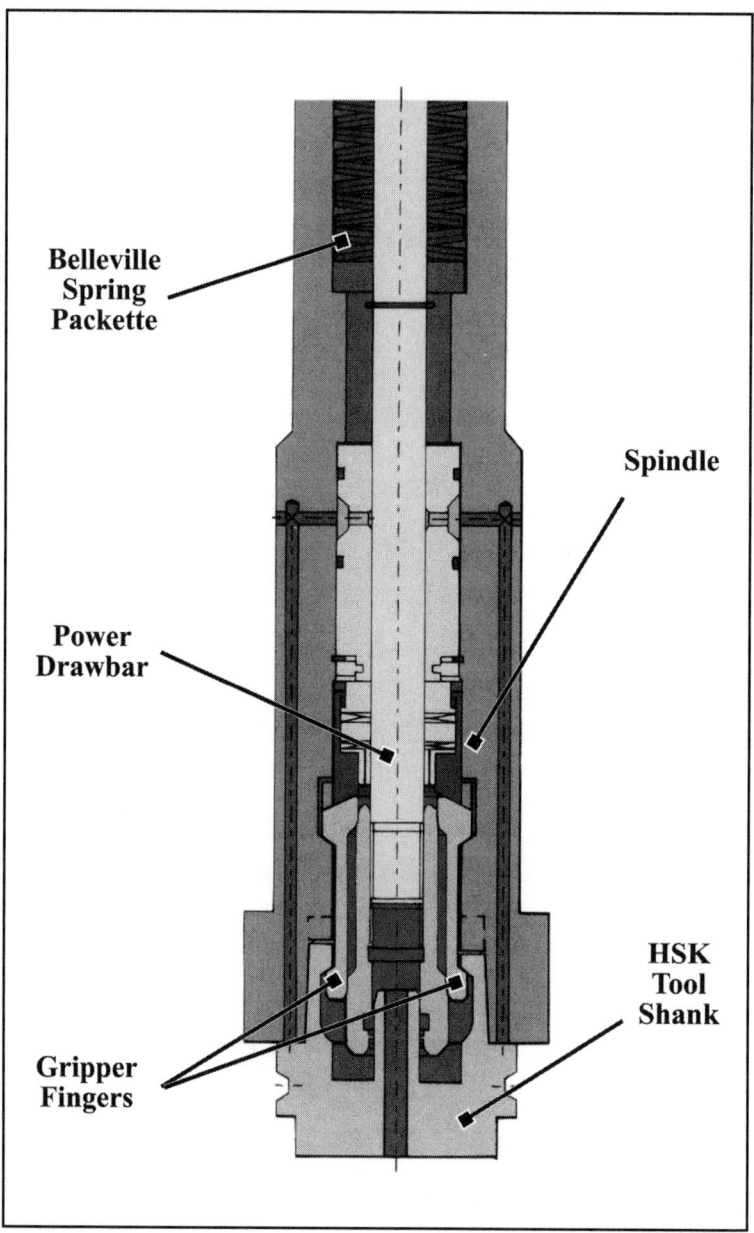

Figure 1.13 • HSK Clamping Chuck for Automatic Tool Change (OTT-System)

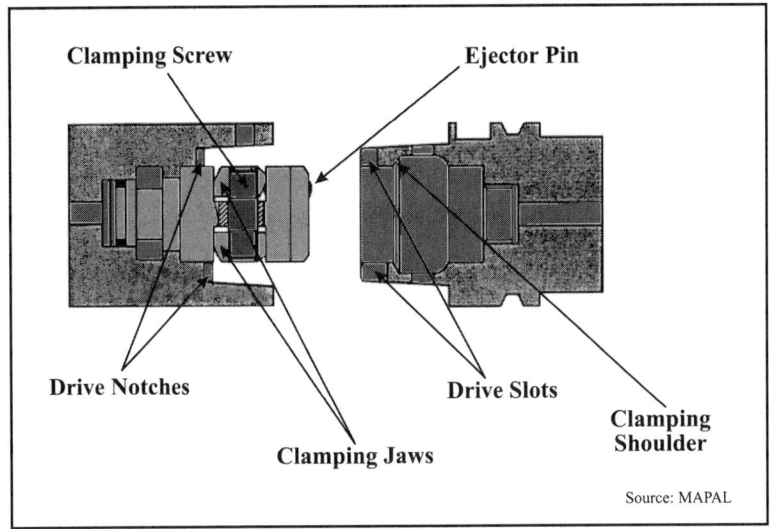

Figure 1.14 • HSK Clamping for Manual Tool Change

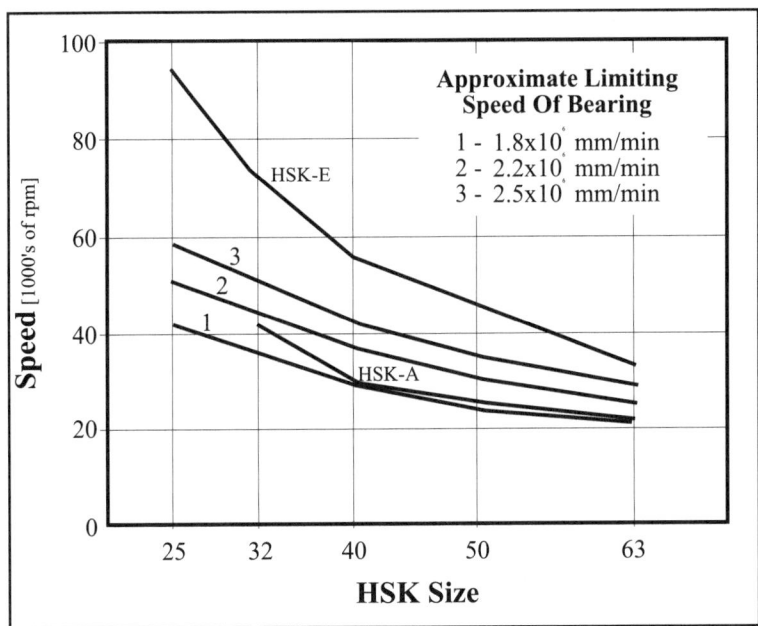

Figure 1.15 • HSK Theoretical Speed Limits

20 • Chapter 1

are other than HSK. In addition, existing machine tools with conventional taper shanks are adaptable to HSK cutting tools and can be applied to the low end of high speed machining. On the other hand, the development of the HSK interface is perfectly timed with the creation of HS/HV machines, so that both are now (and will continue to be) compatible.

CUTTING TOOLS

Chip making machining takes place at the cutting edge, where the machining process is evaluated and measured. The main criteria for cutting tools used for high speed/high velocity machining are cutting speeds and feed rates. These numbers vary greatly depending on the machining operation and workpiece material. High speed cutting can be done on any workpiece material and with any machining operation, but practical limits are dictated by different parameters. See Figure 1.16.

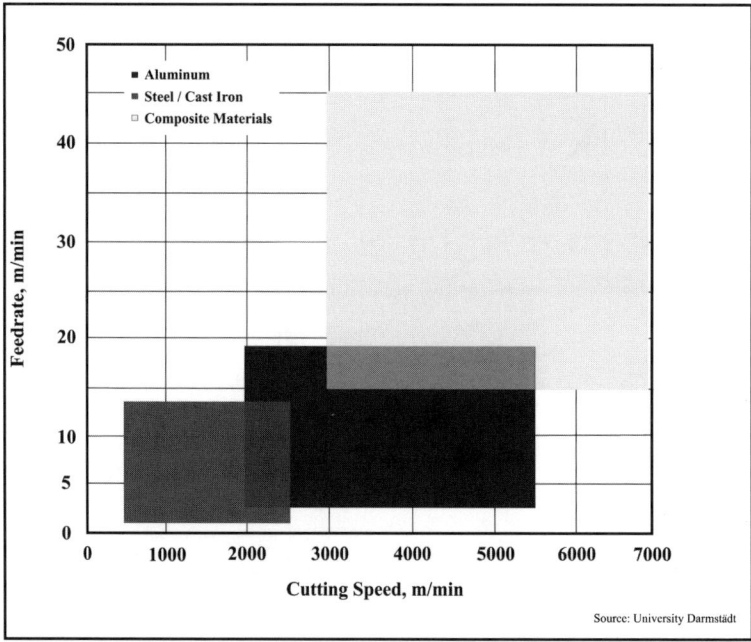

Figure 1.16 • Speed And Feed Ranges For Different Workpiece Materials

Workpiece Material/Cutting Operations

The limits for high cutting speeds are basically set by the workpiece material, the tool cutter material, and the operation itself. Some examples of workpiece material induced limits are as follows.

Workpiece Material	Limits set by
Steel, Cast Iron	Cutter Material
Superalloys	Cutter Geometry
Aluminum, Copper	Spindle Speed
Magnesium	Flammability

Workpiece material and machining have to be tuned to the specific requirements of each application. Machining cast iron and steel, superalloys, aluminum and magnesium, and composites require different techniques.

Cast Iron and Steel

High cutting speeds for cast iron and steel are about 1/3 to 1/5 of that for aluminum. The reason is the thermal breakdown of the cutting edge that occurs due to heat generated during cutting.

Face Milling: More so than with most other materials, the right mix of cutting speed and feed rate is the determining factor for sensible face milling operations. The optimum feed rate is very much dependent on the cutting speed. With conservative feed rates and stock removals that are comparable to traditional milling operations, high cutting speeds of up to 400 m/min can be applied when using advanced cutting materials like cermet, ceramics (Si_3N_4), or cubic boron nitride. Another factor that must be considered is that, at high speeds, the use of coolants has a more negative effect because the cutting tool is subjected to substantial temperature fluctuations that can decrease tool life.

Turning: High cutting speeds are predominantly used for finishing operations to yield good surface geometries. An increase in machining speed has a favorable effect on the workpiece and geometric tolerances, and substantial improvements can be achieved by reducing the amount of stock removal. This also decreases the required cutting forces. In addition, the inherent (albeit somewhat lower) cutting forces, due to higher cutting speeds, will have a positive impact on the surface geometry. Tool life can also be extended under these conditions because the relationship of cutting speeds and feed rates directly affects tool life. Surface finish and accuracy are similar to those of traditional turning. Again, it should be noted that optimal turning is dependent on advanced cutting materials.

HS/HV machining is especially economical in the area of "hardturning," where it replaces grinding operations with the use of cubic boron nitride cutting material at speeds up to 4,000 m/min, using a lathe as a "regular" production machine.

Drilling: Heat buildup, chip formation, and chip disposal can impede workpiece penetration. While heat buildup on the cutting edge and bore wall can be controlled through coolant supply and cutter material, chip control is strictly a matter of drill geometry. The key is to create short, curled chips, shaped like 6's and 9's, that can easily be carried out of the bore through the drill flutes.

If chip volume can be kept slightly lower than with conventional drilling, high cutting speed and feed rates can be applied. This is noteworthy, since drilling is either a "finish operation" of somewhat wider tolerances, or a premachining operation, where the emphasis is more on tool life than surface finish.

Reaming: Usually used for finishing on "regular" production machines, reaming is to be applied at high speeds with low feed rates and minimal stock removal (about 0.5 mm diameter). A combination of cutter material and fine-tuned geometry is necessary for success at high cutting speeds. In a high production environment, cycle time is often as important as finished quality; for that reason, high cutting speeds are used through multi-blade or multiflute designs to complement high feed rates. A favorable side effect is a further decrease of radial cutting forces which reduce the load on the machine and allow the tool to be guided more efficiently in the bore. This, in turn, improves surface finish through the use of high cutting speeds. Reaming usually takes place within a range of 500-700 m/min.

Superalloys

These materials consist of four groups:
1. nickel base (mainly nickel 35-70%, and up to 30% chromium and other materials),
2. cobalt base (cobalt, chromium, nickel, and/or tungsten),
3. iron base (iron, 20% chromium, and 25-35% nickel), and
4. titanium base (titanium, aluminum, vanadium).

All four alloys retain their strength at high temperatures, are highly abrasive, have low thermal conductivity, and tend to work-harden. In other words, they are extremely difficult to machine. The limit for high speed machining is, therefore, set by the cutting material. Machining operations are limited to cutting speeds up to about 500 m/min to obtain economic tool life before the cutter material breaks down. Aluminum-oxide (Al_2O_3), silicon-nitride (Si_3N_4), PCBN (Polycrystalline Cubic Boron Nitride) and, for titanium, PCD (Polycrystalline Diamond) are the cutting materials of choice. Increasing the feed rate will generate extra high cutting forces, and thereby increase the load on the process. Feed rates should start out at 50% of maximum and then they are increased once the tool is "in the cut." Variations with one cutter material, PCBN, when machining Inconel 718 (a nickel base alloy), are shown in Figure 1.17.

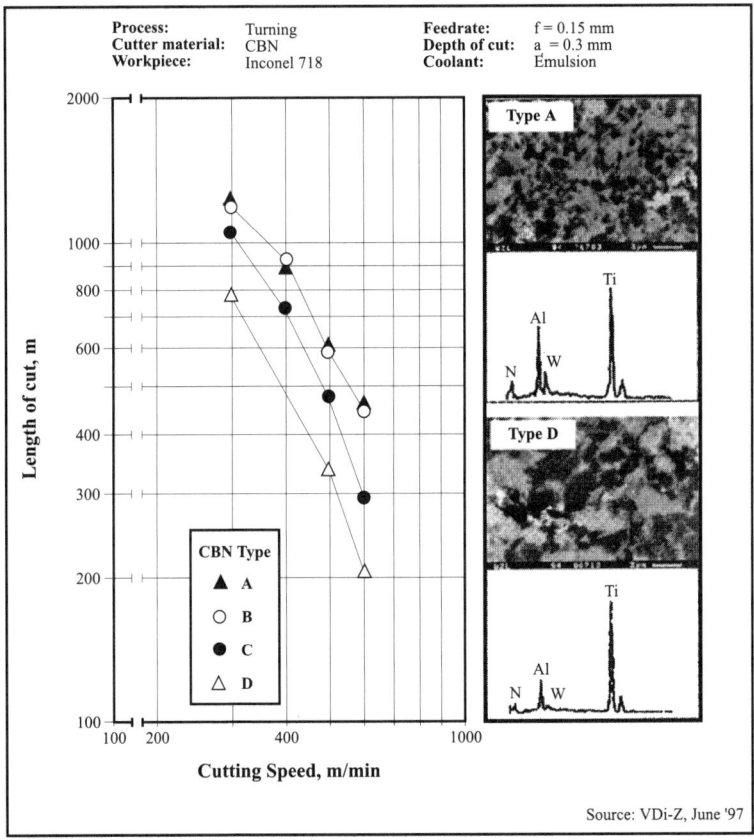

Figure 1.17 • Turning of Inconel 718 with PCBN

All superalloys generate extraordinarily high temperatures during the machining process. This is an added factor at high cutting speeds, especially for titanium alloys. However, because of the high speed, the heat transfer occurs not at the workpiece, but through the chip which might reach the melting point, and that can aid in favorable chip formation.

Aluminum and Magnesium

The need for lightweight construction in a wide variety of industries — notably aerospace and automotive — has put nonferrous workpiece materials in high demand. This holds particularly true for aluminum which also has good machinability working in its favor. One could call it *the* material of choice for machining in an HS/HV environment. It is also the ideal match for diamond cutting tools. Any operation, whether milling, drilling, boring, or turning, can be performed at extremely high cutting speeds and with respectable feed rates. Finish operations use conservative feed rates and provide light chip removal rates, while high speeds help achieve excellent surface finishes. Cutting speeds of up to 7,500 m/min in milling operations are common and, typical for nonferrous machining, low cutting forces result in lower power requirements (65% less than steel, up to 80% less than titanium) as well.

Because of its low density (1.8 kg/dm^3) and its high tensile strength, magnesium is an interesting alternative even to aluminum when extreme low weight is a criterion. Its usefulness for machining is that it offers long tool life and it can be machined with high cutting data, thereby offsetting the higher material cost. The highest applicable cutting speed

Operation	Specific Force [N/mm^2]	
	Aluminum	Magnesium
Facemilling	700	390
Endmilling	180	130

Source: TH Darmstadt

Figure 1.18 • Cutting Forces for Milling Al and Mg

is mandated by the safety of the process: at the critical temperature of 650°C, magnesium chips can catch fire. Rapid chip disposal out of the machining area, and provisions for adequate cooling with special coolants can remedy the problem. Magnesium has all the machining attributes of aluminum and even smaller cutting force requirements, as can be seen in Figure 1.18.

Composites

The two significant groups of composites are graphite-epoxy (carbon-fiber) matrix and metal matrix.

Graphite-epoxy is most commonly machined through forming. Only drilling is performed as a secondary operation. This takes place at very high speeds, which allows the tool to move rapidly, decreases heat buildup, and prevents delamination of the composite.

Metal matrix composites consist of reinforcements (whiskers, fibers) in a metal matrix (aluminum, titanium, copper, magnesium). Some withstand heat buildup better than superalloys, and many can easily be high strength, lightweight replacements for conventional materials. See Figure 1.19.

Material	UTS ksi	UYS ksi	Elongation %	Young's Modulus x 10^6 psi	Fracture Toughness Ksi-in$^{1/2}$	Hardness Ra	Density Lb/in^3
Aluminum Matrix 15% SiCw (0.100 Thick Sheet) Longitudinal Orientation	104	83	5.3	16.5	54	87	0.102
Titanium 6Al 4V	170	160	10	16.5	40 - 60	105	0.162
Aluminum 6061-T6		40	17		27	59	

Figure 1.19 • Physical Properties of Aluminum Matrix Composite Compared to Ti and Al

In order to forego delamination and prevent frazzling of the material, high cutting speeds using mostly PCD have proven to be ideal for most typical machining operations. Depending on the material composition, they can be machined dry or with coolant. Since all composites are "designer materials," they do not limit high speed/high velocity machining performance.

Cutting Material

The following advanced cutting materials are the most suitable:
- Polycrystalline diamond (PCD) for nonferrous metals,
- CVD-diamond coating for nonferrous metals,
- Polycrystalline cubic boron nitride (PCBN) for ferrous metals,
- Ceramics (Al_2O_3, Si_3N_4) for ferrous metals, and
- Cermets for ferrous metals.

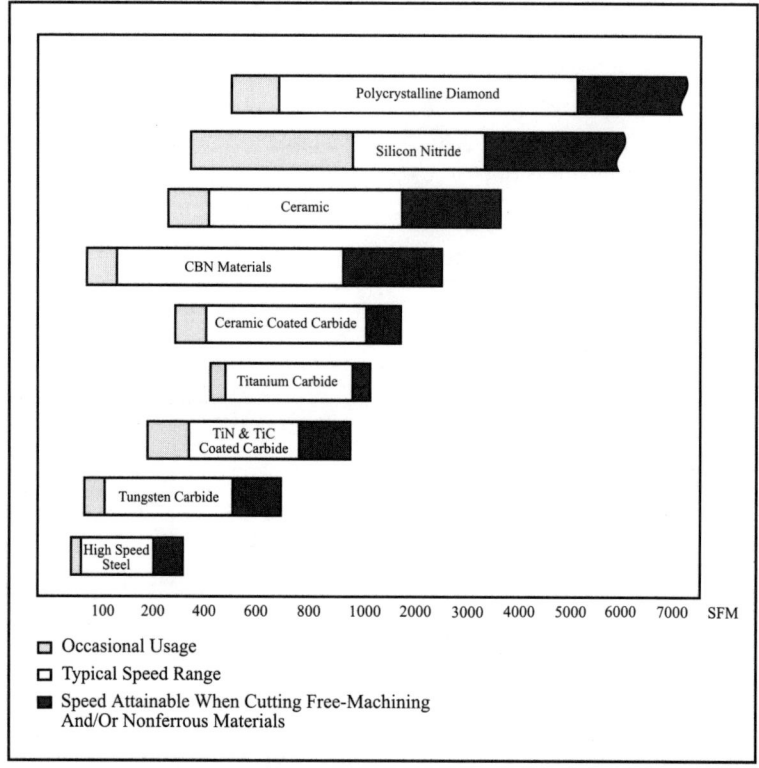

Figure 1.20 • Speed Ranges for Various Cutting Tool Material(s)

The high end of high cutting rates would be at about 2000 sfm for CBN and cermet in ferrous metals, while this would represent the low end for PCD in nonferrous metals. See Figure 1.20.

The hardness of the cutting tool is more important than toughness when using high cutting speeds. As speed increases, the qualities of hardness, abrasion resistance, chemical inertia, and matrix interaction become even more important as is shown in Figure 1.21.

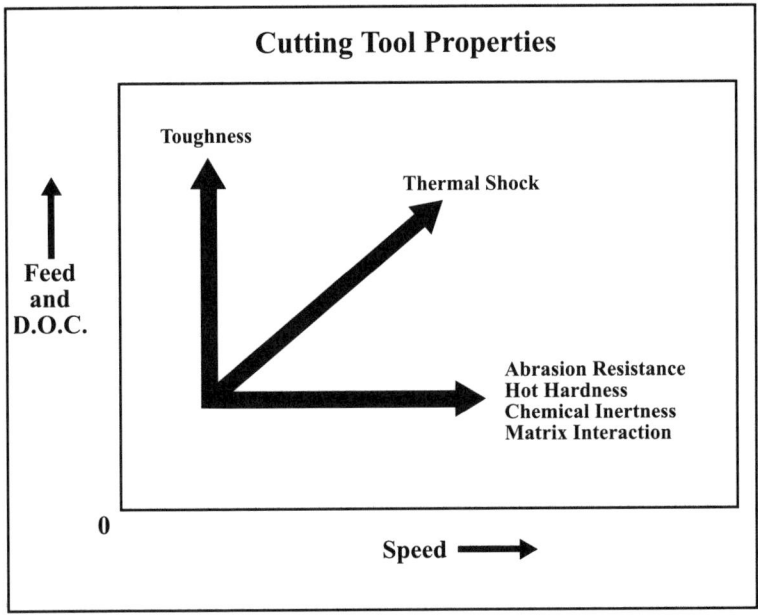

Figure 1.21 • Cutting Tool Properties

Polycrystalline Diamond: PCD is an extremely hard and wear-resistant cutting material. It can run at virtually unlimited cutting speeds in any soft nonferrous metal. Even in abrasive aluminum alloys with high silicon content between 12 and 18% (Al 390), it can be machined within runs of 5 digits. Since higher speeds produce better surface finishes, it is the ideal material for finishing passes. High metal removal rates and long tool life also accommodate the one-pass machining principle. Its limitation is not so much the rotational axis with which a part is machined, but at what point the part might move out of place, the grade of cut, or the system's balance. PCD, run wet or dry, constitutes a true breakthrough in cutting technology. (See Figures 1.22 and 1.23 for hardness and abrasion resistance values.)

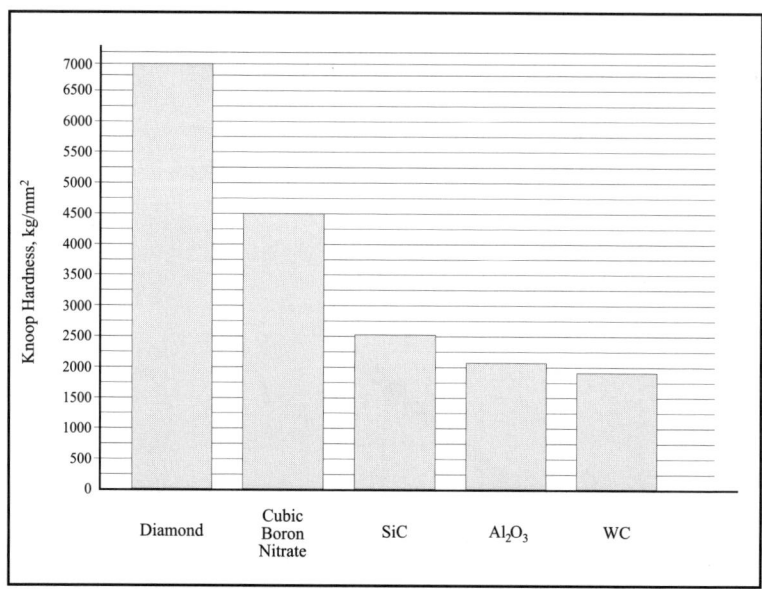

Figure 1.22 • Relative Knoop Hardness

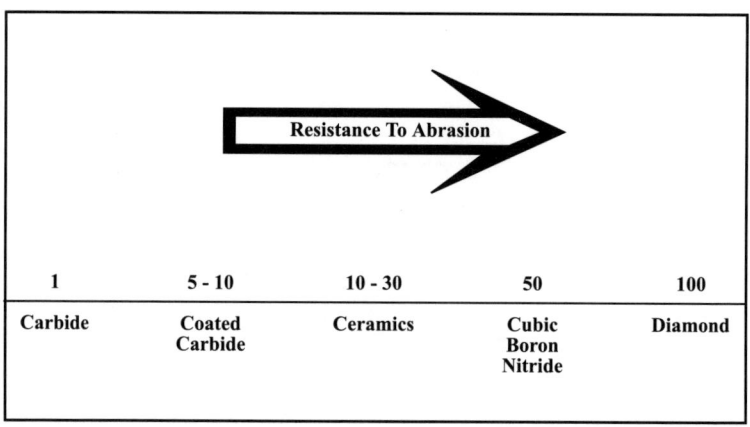

Figure 1.23 • Range of Abrasion Resistance

CVD Coating: Efforts to find less expensive alternatives to polycrystalline diamond have been under way for more than a decade. A viable alternative for some applications is CVD coating. However, when CVD layers are "grown" on a carbide substrate, the adhesion is relatively poor because of the cobalt content. On a ceramic substrate, there is no adhesion problem, but the material becomes too brittle and has extremely low shock absorption.

Today's diamond coated carbide (DCC) can be shaped more easily for any cutting geometry than PCD; it has about the same hardness as PCD, and it can run at high speeds. However, its tool life is lower because the layers wear off faster, and its cost equals or exceeds today's PCD. Diamond coated ceramics can be successfully used only when there are no radial loads on the tool, and when it is operating under very stable, lower speed applications. The conclusion is that the development of DCC tooling is promising. But, unless it can economically outperform PCD, the latter is and will remain the more widespread cutting material.

Polycrystalline Cubic Boron Nitride (PCBN): Its high hardness, wear, and temperature resistance enable PCBN to machine hard ferrous alloys such as hardened steel, superalloys, hard gray cast iron and powdered metals. It is suitable for interrupted and continuous operations, and in "hard turning," it is a good replacement for grinding operations.

Material	Operation	PCBN Grade	Surface Speed ft/min	Feed Rate in/rev	Depth of Cut in
Gray Cast Iron (180-270 BHN)	Turning Milling	BZN-6000 BZN-6000	2000 - 4000 2000 - 4000	.006 - .025 .006 - .012 (in/tooth)	.005 - .100 .010 - .100
Hard Cast Iron (>400 BHN)	Turning Milling	BZN-6000 BZN-6000	250 - 500 400 - 800	.006 - .025 .006 - .012 (in/tooth)	.005 - .100 .010 - .100
Hardened Steel (>45Rc)	Rough Turning Finish Turning Milling	BZN-6000/-8100 BZN-8100 BZN-6000	220 - 350 350 - 450 400 - 800	.006 - .025 .004 - .008 .004 - .010 (in/tooth)	.030 - .100 .004 - .030 .004 - .075
Superalloys (>35Rc)	Turning Milling	BZN-6000 BZN-6000	550 - 800 700 - 1000	.004 - .012 .004 - .008 (in/tooth)	.004 - .100 .004 - .050
Sintered Iron	Turning Milling	BZN-6000 BZN-6000	300 - 600 400 - 800	.004 - .010 .004 - .008 (in/tooth)	.004 - .050 .004 - .050

Figure 1.24 • PCBN Operation Guidelines

The key to choosing the right machining data is to find the optimum machining parameters.

Optimizing machining data is a matter of cutting speed and feed rate versus cost. This aspect is especially important for difficult-to-machine workpieces because, with increasing cutting speeds, cost advantages are obtainable through high productivity by reducing main machining time. This also entails higher costs through higher tool wear and increased noncutting time. An increase of feed rates reduces the time the tool is in the cut, and thus reduces the cost of machining. But, again, too high a feed rate can cause premature tool failure. See Figure 1.25.

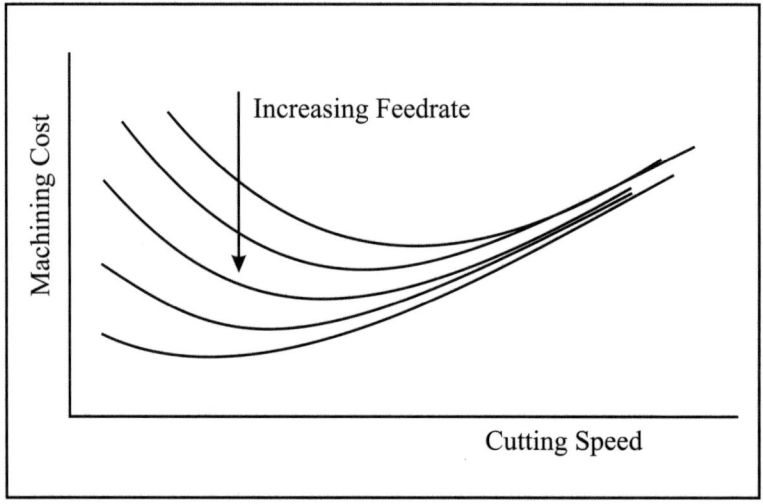

Figure 1.25 • Machining Cost

Ceramics: Basically there are two different kinds of ceramics — aluminum oxide based (Al_2O_3) and silicon-nitride based (Si_3N_4). Ceramics are especially suited for gray cast irons at high cutting speeds because of their hot hardness without plastic deformation. Their brittleness, however, makes them susceptible to mechanical shock. Si_3N_4 is recommended for high-speed roughing and, because of its relative toughness, can also cut interruptions, while Al_2O_3 is for high-speed finishing (see Figure 1.26). For "hard turning," ceramics constitute an interesting alternative, and for softer cast irons it can be attractive due to cost considerations.

Cermets: These materials consist of titanium carbide and titanium nitride with a binder such as tungsten, molybdenum, nickel, cobalt, or tantalum. They are primarily used for machining steel at higher cutting speeds and conventional feed rates. They feature favorable friction behavior and low susceptibility to wear. They are also an economic cutting

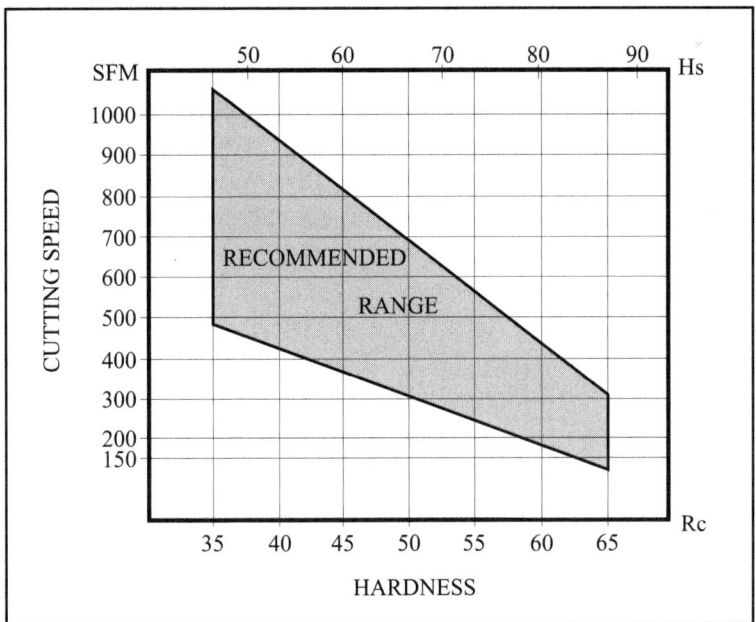

Figure 1.26 • Ceramics - Range Of Cutting Speeds

material that deserves to be used more often for fine-finish turning and reaming operations, where they achieve respectable tool life at cutting speeds beyond any carbide grade.

Special Design Features

When machining at extremely high speed and high acceleration rates, special cutting design and engineering characteristics have to be taken into account. The most relevant are: cutter geometries, tool support, tool overhang, adjustability, coolant, and chip control. Fast removal rates usually preempt lighter cuts with the emphasis more on cutting speed than on feeds and the stock removal (depth of cut) is usually somewhat less than normal.

There are some fundamental principles of insert technology. For example, the fact that the strength of the insert is determined by its shape and its cutting edge, and is highest with negative cutting edges, is the same as with conventional and HS/HV cutting conditions. Some geometries simply have to be adjusted to different cutting situations. Lead angle, chip breaker, and cutter edge are of special importance. Basically,

32 • Chapter 1

clearance angles should be kept to a minimum for edge strength. Chipbreakers, in order to provide proper chip formation, should have a cutter edge in the form of a small radius to lessen the load (cutting force) and provide good surface for finish operations. A larger radius should be provided to strengthen the cutting edge for machining hard metals (superalloys, alloys).

Figure 1.27 depicts an advanced geometry well suited for high speed fineboring, featuring a primary and secondary cutting edge with a ground-in 1° relief angle. This design semi-finishes and finishes during one cut.

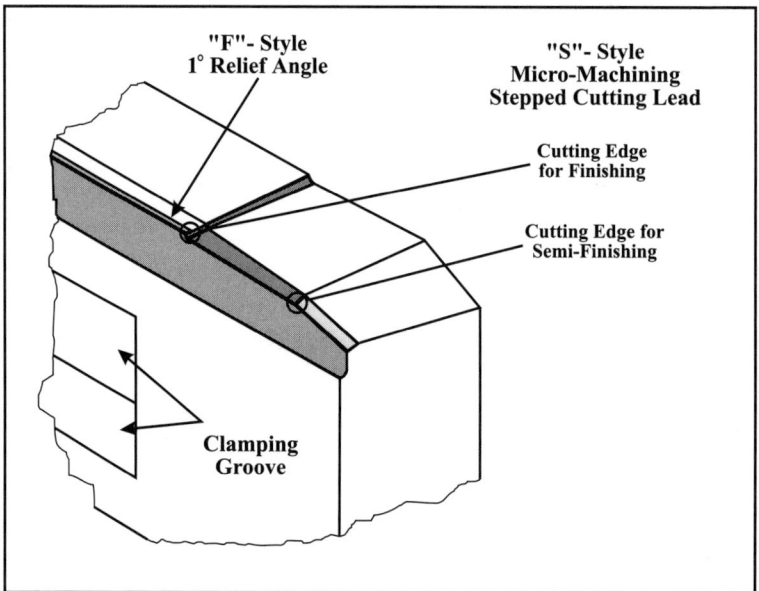

Figure 1.27 • Advanced Cutting Geometry

Often, optimum insert geometries and edge preparations have to be determined empirically through tests, simply because of the wide mix of machining characteristics and the lack of case histories of removing a specific material at fast rates. This can be a time-consuming proposition, but is still better than optimizing a process during regular production runs.

One of the keys to successful tool design in HS/HV machining is to support and stabilize the tool right at the beginning of the cut and maintain it throughout. How it is done depends on the tool style and the operation. The individual machining operation predetermines such a design. For milling it could be the shape of the flutes; for turning the cutting angles; for face milling the arrangement of roughers and wipers; for boring peripherally opposed cutters; for multiflute reaming ground-in lands at the trailing end of the cutting edges or just four instead of two flutes; and for reaming with indexable blades, peripherally arranged guide pads that axially slightly trail the cutting edge. Besides these design features for straight, aligned cutting, rotation and acceleration forces ought to be applied at the low end as the cutter begins machining and then, as it stabilizes itself, be taken to their respective maximum so as to avoid mechanical shock and cutter deflection.

This leads to another factor that has to be considered — tool overhang. Figure 1.28 shows this factor that determines tool deflection. It is well known that the length of the tool is very influential, but the equation also interprets the positive effect of both the tool's mass and a higher modulus of elasticity, i.e. dense tool material, which is often overlooked.

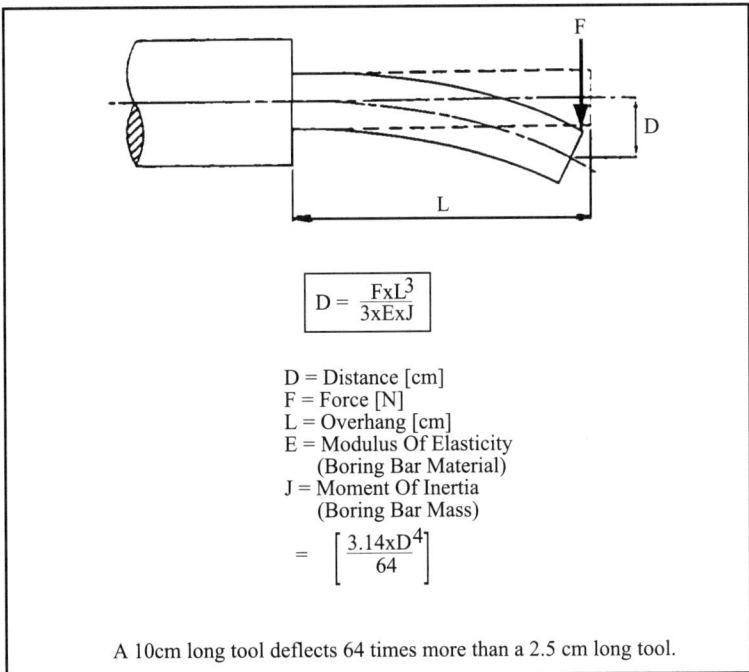

$$D = \frac{FxL^3}{3xExJ}$$

D = Distance [cm]
F = Force [N]
L = Overhang [cm]
E = Modulus Of Elasticity
  (Boring Bar Material)
J = Moment Of Inertia
  (Boring Bar Mass)
  $= \left[\frac{3.14xD^4}{64}\right]$

A 10cm long tool deflects 64 times more than a 2.5 cm long tool.

Figure 1.28 • Tool Deflection

Heavy metal, consisting of tungsten (about 90%), nickel (about 3%) and iron (about 7%), has a much higher modulus of elasticity than regular tool steel. Therefore, it inherently dampens the cutting process and aids in chatter-free machining of critical finish operations.

Adjustability: If an unstable machining process (caused, for instance, by a worn spindle or an unfavorable misalignment of the tool) prevails, the process does not necessarily have to be aborted. The provision of an inexpensive adapter, that can be trammed in to eliminate axial or radial misalignment between machine spindle and tool, can save an otherwise doomed cutting condition.

When machining has to be done with coolant supply, its flow has to be uninterrupted and lead directly to the cutting edge with the right pressure and volume. This is to cool and/or lubricate the process and, particularly important for high cutting speeds, to discharge the chips out of the immediate cutting area. The above factors are an essential and equal part of the high speed/high velocity equation.

Applied Tool Management

Mikio Kitano, President of Toyota-America, has said that future manufacturing will stress and execute the principle that "smaller and lighter is better." I would like to extend this philosophy to "smaller, lighter, and faster is better — but not easier." Technology will make the shaping of parts and components more complex, and will make the production floor more disciplined in executing customers' orders. Part of the technology will be defined and applied tool management. It will be irrelevant whether the tool management is done in-house or by a third party (Tier 1-, Tier 2- supplier concept). With the advent of flexible manufacturing about two decades ago, it became clear that the vast amount of tooling that individual machines were loaded with had to be intelligently managed. This basically meant logistics: having enough tools available at any time, preset and ready to be loaded into the machines' magazines. This is where it essentially ended and where it remains today.

"Smaller" and "lighter" workpieces imply that more functionality is being designed into more thin-walled parts of more expensive material. To machine them "faster," in turn, implies that there is a lot less room for error — no window for mistakes. That is where the concept of *Applied Tool Management* comes in. When a part is machined, the machining data have to be monitored to establish the point where adjustments need to be made in advance. Consider the following example. A fineboring tool with a diameter of 20 mm, held in an HSK-F holder, is finishing an aluminum bore at 29,000 rpm and 1 mm/rev with a 1 mm DOC. After machining 10,000 parts, the C_{pk} requirement indicates that the parts deviate from the range. The power requirement of the machine is still the same, so the out-of-tolerance condition needs to be passed on to tool management, who can set the tool to make it stay within the desired tolerance range. Only timely and precise feedback can secure correct machining and contribute to predictable results.

Technology for rapid manufacturing is expensive. Fast, continuous fine-tuning of cutting tools is a pivotal part of advanced tool management. It is this function that allows machining with fast removal rates to go full circle.

The Level of Machine Tool Technology

Traditionally, the machine tool and cutting tool industries have often inspired each other through technological leapfrogging. Machining at high speeds is a case in point. Following the demand by machine tool builders for more reliable and longer lasting cutting material for use with nonferrous metals, PCD cutting material was born. Its capabilities inspired the machine tool industry to develop machines with higher spindle speeds, which opened up avenues for the new HSK interface. The key is for one industry to develop new state-of-the-art technology based on developments by the other.

Multipurpose Tooling: The chance for high productivity and simultaneous precision is, to a great extent, the direct result of advanced tooling systems. A good example is the use of circular milling, shown in Figure 1.29. In circular milling, precision-made fixed PCD tools with HSK interface take up the speed and accuracy of the machine's control system for face milling an aluminum transmission case at very high cutting speeds and feed rates. To accommodate the all-important time element, processes have to be laid out that call for the minimal number of tools and shortest path for the cut. Circular interpolation, the use of multipurpose tools, and the one-pass machining concept are to be specified wherever possible (see Chapter 3).

Balancing Safety: Because the centrifugal force increases as the square to the rotational velocity at high speeds, the balancing of the tooling system and the design of the tools has become an issue.

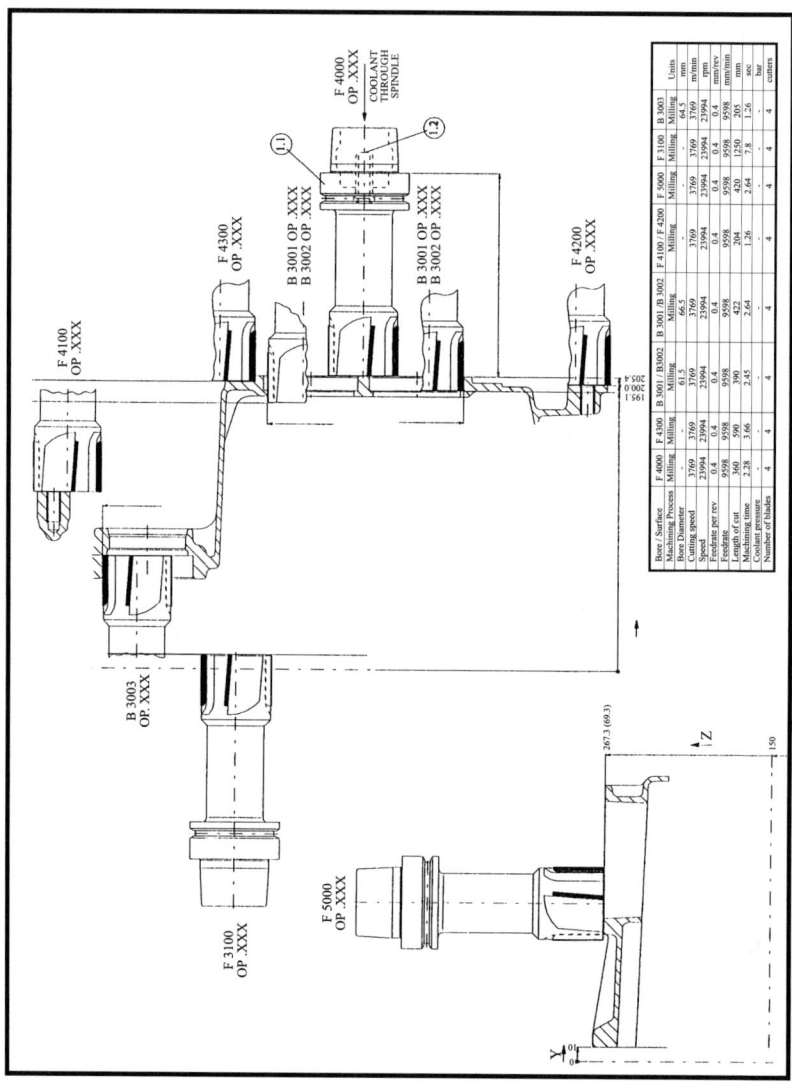

Figure 1.29 • Circular Milling Tools

Excessive imbalance of the tooling system has an effect on the workpiece (chatter marks, out of tolerance) and the machine (premature spindle failure). In worst-case scenarios, imbalance causes a tooling system to cut "out-of-control," break or damage workpieces and/or the machine, and become a safety hazard for machine operators. To avoid an out-of-balance tool, both toolholder and machine need to be balanced as close to the same balancing grade as possible. However, good balancing notwithstanding, sheer centrifugal force alone is cause for concern, and cutting tools have to be designed to secure safe operations.

Tool failures can be caused by disintegration of the tool body itself, by lifting the cutter clamp plates, by bursting of the cutting inserts, or by deformation of connecting hardware (cartridges, bolts, etc.). Design layouts have to be based on mathematical configurations that include the strength to resist centrifugal forces, and then need to be enhanced through experimental tests to determine at what elevated speeds parts start to move, why they begin to move, at what speed they eventually break apart, and when catastrophic destruction of the tool occurs.

Design priorities include:
- minimizing the tool's mass,
- striving for tool symmetry,
- building the tool with smooth surfaces,
- use of a minimal amount of hardware,
- providing form and pressure clamping for inserts,
- eliminating frequent tool adjustments,
- having smaller elements fail first in case a failure does occur, in order to lower its impact, and
- eliminating high stress areas.

Following mathematical equations and design characteristics, the calculations must be verified through tests that can empirically determine the operation range based on given rpm's. The example shown in Figure 1.30 (b) depicts a face milling cutter with the corresponding operation ranges with a built-in 2.5 safety factor.

Optimum Cutting Speed

Practically all cutting materials can be applied within a wide range of cutting speeds, feeds, and depth of cut. The chosen data have to comply with the required surface geometries, the performance of the machine spindle, and the stability of the whole machining process in relation to

Figure 1.30 • (a) Face Milling Cutter

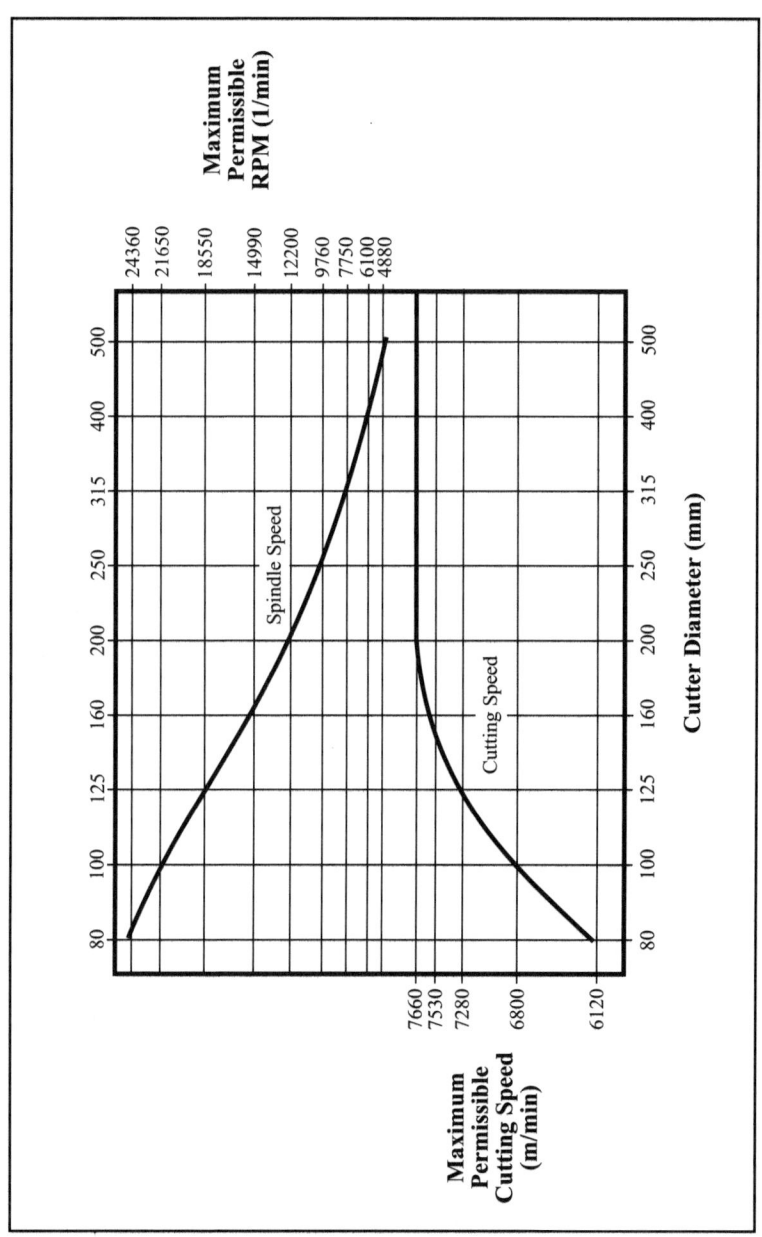

Figure 1.30 • (b) Permissible Cutting Speed — Face Milling

the mechanical and physical characteristics of the material. In addition, cost considerations of the cutting performance *vis-à-vis* the wear of the cutter have to be taken into account.

As previously mentioned, cutting speed plays a predominant role in the condition of the cutter and ultimately its life span (see the "Taylor" curve illustrated in Figure 1.31).

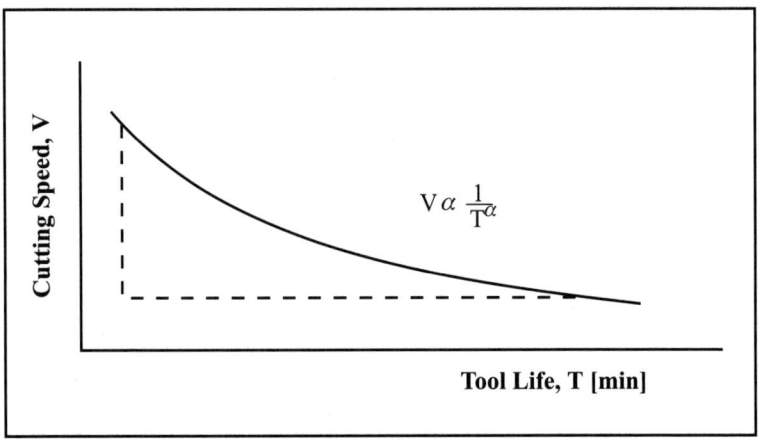

Figure 1.31 • "Taylor" curve

The true goal on any production floor is to minimize the cost per part and maximize its quality, and cutting data are to be chosen accordingly. The challenge is to find the balance between counteracting influences. On the one hand, higher cutting speed can increase productivity by lowering the main machining time. This reduces the cost of both the machine and the operator. On the other hand, increased cutting speed lowers tool life which leads to more frequent tool changes, and increased nonmachining time and other indirect tooling costs (tool management, etc.).

In order to find the optimum cutting speed at which all costs of the process are in balance with each another, the relationship of cutting speed and machining cost must be studied and contrasted, as illustrated in Figure 1.32.

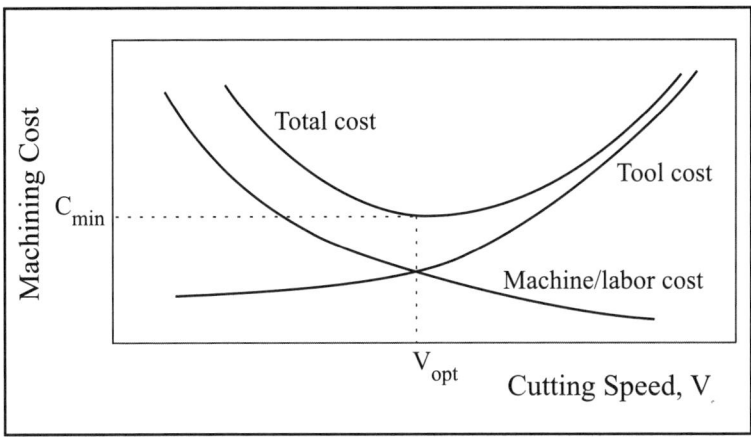

Figure 1.32 • Figuring Optimum Cutting Speed

Conclusion

High speed/high velocity machining can be a formidable manufacturing technology if approached sensibly — with the machine, tooling interface, and cutting tool all being given equal weight. HS/HV machining can only fulfill its promises if all individual aspects in this technology chain are designed, engineered, manufactured, and applied properly.

Machine technology has to be used to maximize tooling technology in order to compose an advanced manufacturing process, because it is the process, and not the individual components, by which parts are manufactured. The parts are, in turn, the final yardstick that a manufacturing company is measured by in terms of productivity, quality, and economics — the measures of failure or success in the marketplace.

Chapter 2

DRY-MACHINING AND NEAR DRY-MACHINING

Every year, the manufacturing industry consumes millions of gallons of machining coolants. While these coolants aid in the efficiency of many machining processes, they are costly and represent a potential threat to operator health and to the environment. Cost, as well as health and environmental issues, mandate that manufacturing companies strive to drastically reduce coolant consumption and, if possible, eliminate it altogether.

Recent studies have found that the cost for acquiring, maintaining, and disposing of coolants represents between 7% and 16% of total production costs (see Figure 2.1). Cutting tools, by comparison, stay well within single digits on any manufacturing process. Furthermore, health experts and environmental groups point out that coolants contribute to the contamination of soil (leakage), water (disposal), and air (airborne particles), and can be linked to skin and lung diseases. Manufacturing, always under pressure to lower costs and to employ environmentally friendly materials, urgently needs to find competitive alternatives to traditional "wet"-machining.

Why have we held on to coolants, if there are so many repercussions? Quite simply, coolants perform three essential functions: they cool and lubricate the machining process and they aid in chip

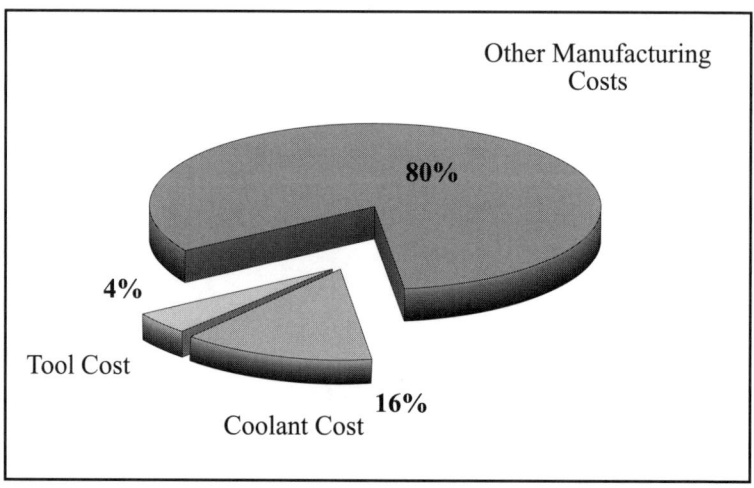

Figure 2.1 • Percentage of Cost for Coolant and Tools as Part of the Total Manufacturing Cost

discharge. In fact, during machining, coolants have a profound and positive impact on the workpiece, the cutting tool, and the machine. The following list presents the various benefits provided by coolants in manufacturing operations:

Workpiece	Cutting Tool	Machine
Geometry	Tool Life	Chip Discharge
Tolerances	Adhesion	Temperature
Thermal Stress	Diffusion	Corrosion
Surface Finish	Chip Formation	Cleaning
Cleaning	Heat Buildup	
Corrosion Protection	Chip Discharge	

A substitute for these vital, positive effects cannot be found by simply turning off the flow of coolant. Instead, today's technological wisdom tells us that we will be hard-pressed to ever eliminate "wet"-machining across the board. However, using the knowledge and technology available to us, we do have an alternative: the road to dry- and near-dry-machining is very promising.

Before discussing this emerging technology, we need to identify the major machining operations and group them into geometrically defined and nongeometrically defined processes, as shown in Figure 2.2.

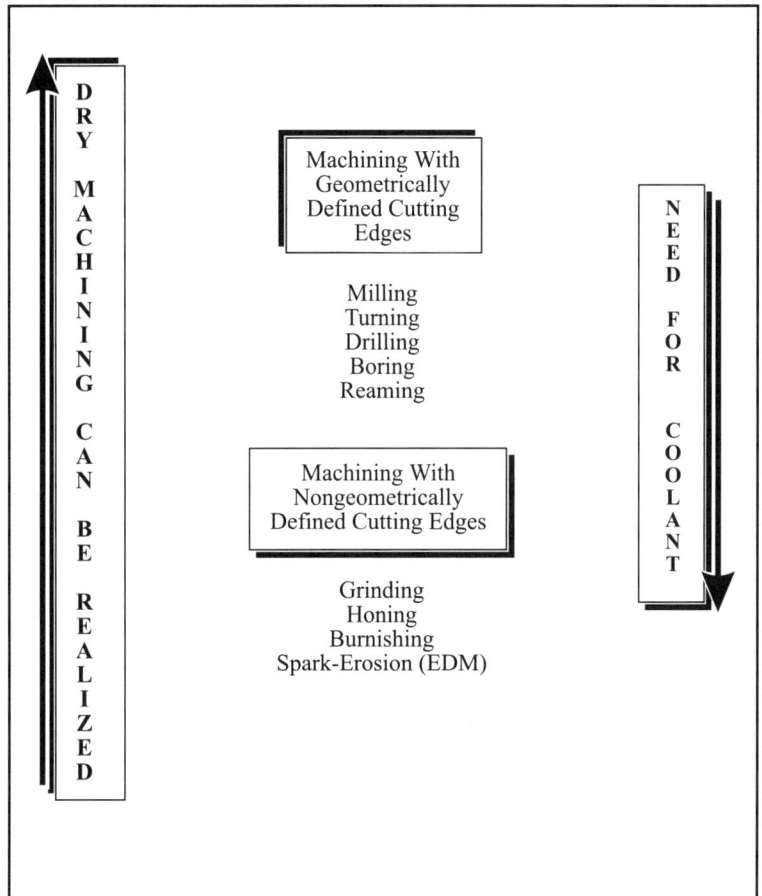

Figure 2.2 • Machining Operations and Their Need for Coolant

Non-geometrically defined machining processes (grinding, honing, burnishing, EDM) are very coolant sensitive. When using these processes the key to reducing coolant use should be to exercise restraint, optimize consumption and, more importantly, use substitute processes out of the group of geometrically defined machining processes whenever possible. With these processes, successful machining without fluids can be achieved.

46 • Chapter 2

Dry-Machining

Eliminating coolants — and with them their primary functions (cooling, lubricating, and chip flushing) — has consequences for the workpiece, the cutting material, and the machine. In addition, the machining process will produce much higher friction and more adhesion between the workpiece and the tool. The heat generated and transferred to the workpiece, cutting tool, and chip is no longer absorbed and taken out of the immediate cutting area by the coolant (see Figure 2.3). The associated thermal stress on the workpiece, tool, and machine, and the lack of lubrication during cutting, have to be dealt with through innovative design and new technologies.

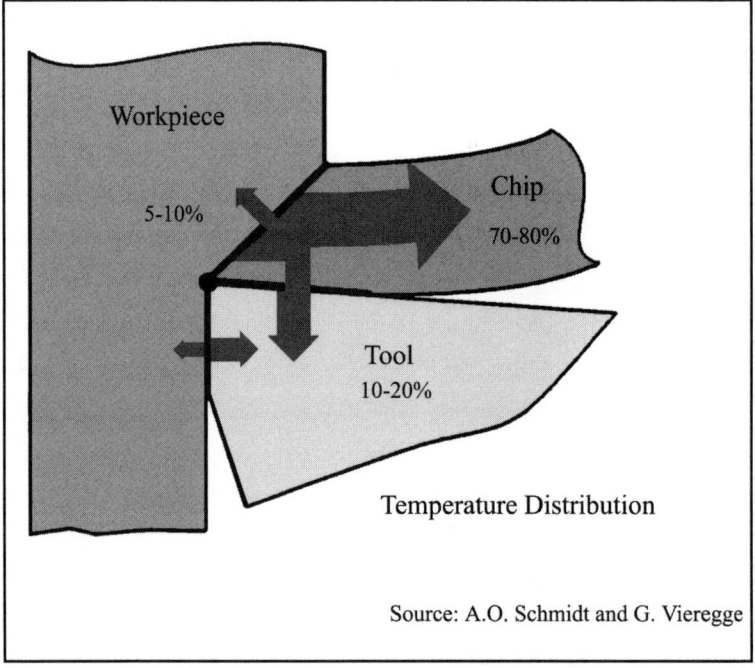

Figure 2.3 • Transfer of Heat Generated During Cutting

Cutting Material

Dry-machining requires that the cutting material's possess extreme hot hardness and toughness in addition to abrasion-, thermoshock-, and adhesion-resistance. This specified profile, due to complex mechanical and thermal load conditions, can be accommodated with advanced cutting materials specifically coated for the machining task. The comparative hardness and tensile strength of common cutting materials is illustrated in Figures 2.4 and 2.5.

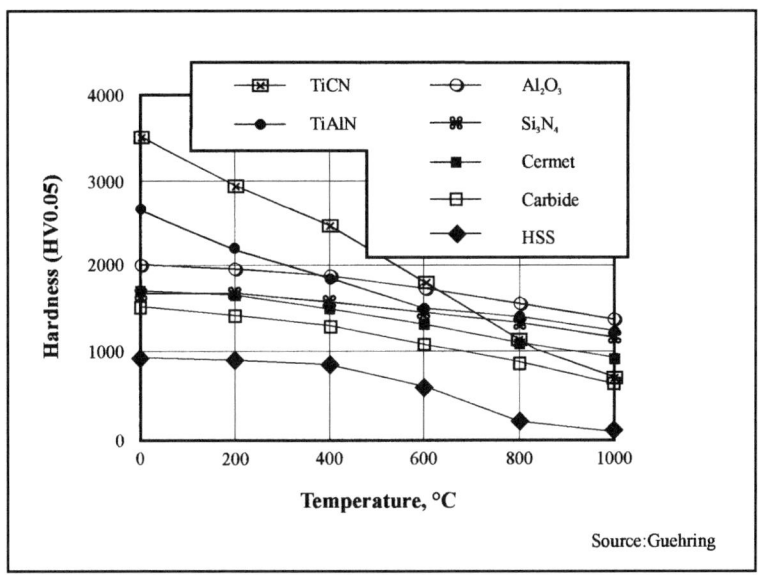

Figure 2.4 • Hardness of Cutting Material

Because of the lack of coolants, dry-machining often subjects the process to elevated cutting temperatures for which hot hardness is required. The ceramics Al_2O_3 (aluminum-oxide) and Si_3N_4 (silicon-nitride) as well as cermets, because of their high hot hardness, are well suited for general-purpose machining without coolants. However, due to their lower toughness, they are more brittle and should not be used for interrupted cuts. As a result, ceramics are good for turning, but are not suited for milling.

HSS (high speed steel) and carbide are usually not suited for dry-machining since both lack the necessary hot hardness, with the exception of fine-grain carbide which has a hardness level comparable to cermet. HSS, carbide, and cermet become suitable for dry-machining through PVD (physical vapor deposition) coatings of TiCN (titanium carbonitride) and TiAlN (titanium aluminum nitride), with the latter hav-

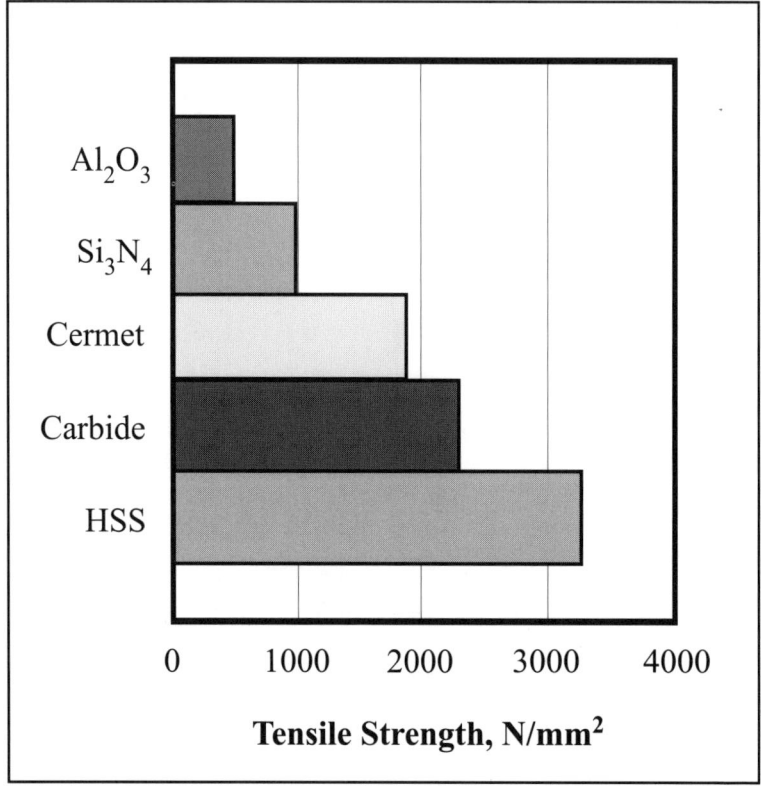

Figure 2.5 • Tensile Strength of Cutting Material

ing more hardness at extreme high cutting temperatures. These coatings assume the dual function of cooling and lubricating the machining process by providing a layer to protect the substrate from thermal or chemical failure.

The newly developed coating MoS_2 (molybdenum disulfide) is a soft coating, featuring a friction coefficient of about 0.01 (one digit less than TiN). HSS PVD-coated tools with a thin layer of MoS_2 yield good surface finishes together with good life in milling and some drilling operations. Interesting also is the combination of the two coatings TiAlN and MoS_2 to a multilayer coating to simultaneously achieve both high wear resistance and low friction. MoS_2, it should be noted, was primarily developed to add to CVD (chemical vapor disposition) diamond coatings as the cutting material of choice for machining nonferrous metals. Another coating called DLC (diamond-like carbide), an amorphone carbon layer with hydrogen and tungsten carbide-like CVD coating, offers an excellent antiadhesion effect.

Even for CBN (cubic boron nitride), the ideal material for dry cast iron machining, coatings are being developed in order to increase toughness for machining a wider range of materials including steels, superalloys, and hypereutectic aluminum.

Tool Design and Machining Data

The focus for optimizing tool design for dry-machining is on cutting geometry and chip removal. For instance, a reduction of the cutting force through a wider clearance angle yields less tool stress and lowers the cutting temperature. Defined rake angles aid in breaking the chips. Wide chip galleys and special flute design (drilling and threading) assure fast and easy exit of cutting chips. Basically, most other design features are similar to machining with coolants, except for specific tasks and operations.

The area of machining data is more complex. In dry-machining, the data usually have to be tuned to variables such as the workpiece, the machine, and the individual operation. However, there are certain guidelines to follow regardless of process and application.

Fact: High cutting speeds are generally desirable for dry-machining. They lower both the tangential cutting force and the workpiece temperature because the cutting tool and workpiece are in contact for shorter periods during cutting. The lower the workpiece temperature, the less surface distortion, and the more likely that parts will be finished to blueprint specifications.

Fact: High feed rates shorten the cutting chip length and minimize the cutting heat that must be absorbed by the workpiece. Instead, much of the generated heat is absorbed by the cutting chips. Unfortunately, higher feed rates also require higher cutting forces.

Fact: Higher depths of cut (DOC) result in increased cutting forces and more stress on the cutting tool, which can lead to premature tool failure even though it might not increase the workpiece temperature disproportionally. If a workpiece needs to be machined at high DOC, it should be done in two or three passes (pre-, semi-, and finishing), possibly allowing for short cooling-down periods between passes. Lighter cuts can be done in one pass.

50 • Chapter 2

Workpiece

Given that the technology is now available to perform dry-machining, the workpiece often holds the key to deciding whether to use wet or dry, since the quality of the finished part is an important element of the process.

The general trend in workpiece design is toward lighter, smaller and as (near) net shaped as possible. Considering that in dry-machining, up to 10% of the cutting temperature escapes into the workpiece, this can have a detrimental effect on geometric tolerances and surface finishes.

Ideally, it is desirable to transfer as little heat as possible into the part. As previously mentioned, this can be done through the right choice of machining parameters. It has been proven that higher cutting data, measured in volume per time unit, can reduce the workpiece temperature and thus restrict heat absorption.

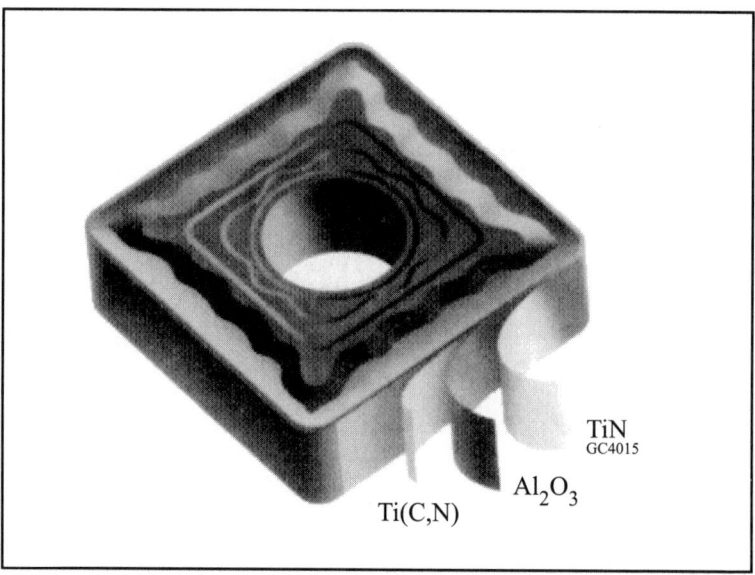

Figure 2.6 • Heat Protective Coatings of Carbide Inserts

The thermomechanical characteristics of the workpiece will vary from material to material. Of course, it has been proven that a great part mass is more suited for dry-machining because it will warm up less easily than a small mass. Dry-machining requires a material with a high temperature capacity and low temperature conductivity. Aluminum, for example, has a high temperature conductivity and absorbs more heat during the cutting process — thereby promoting workpiece deformation due to heat expansion. This makes it a difficult material to cut without coolants.

(Gray) cast iron is on the other side of the spectrum. Its small heat expansion coefficient, low cutting temperatures, and short chips (higher melting point compared to aluminum) make it a better choice for dry-machining.

To pursue dry-machining from the workpiece perspective, the following criteria must be considered:
- increased metal removal rate per time unit,
- the use of tool geometry that reduces cutting force,
- workpiece density and heat conductivity, and
- the machining operation.

The most challenging workpiece/operation combinations are shown in Table 2.1.

Workpiece Material	Machining Operations				
	Turning	Milling	Reaming	Threading	Drilling
Cast Iron					
Steel Alloys		*	*	*	
Hardened Steel		*	*	*	
Aluminum		*		*	
Superalloys	*	*	*	*	*
Composites					

Table 2.1 • Challenging Workpiece/Machining Operations

Machine Tool

Removing coolants from the machining paradigm requires special design features for the machine tools. The two areas affecting the machine are thermostress and chip control. Existing machines will have to be modified, and new ones have to be designed accordingly.

The key is to make provisions inside the machining compartment for easy, fast, and reliable handling of cutting chips. This can include a vertically arranged spindle for the natural drop of chips through gravity. Angled sheet metal with a slant bed design can be used to direct chip collection; and insulated sheet metal can be positioned to prevent heat transfer to vital machine parts. Compressed air (200-300 psi) directed at the workpiece table and the part fixture will free the machining area from chip buildup. However, it might also penetrate air gaps between cover sheets. A defined air suction system is more costly, but can be a better choice.

A closed-loop, built-in cooling system will contribute to the thermal stability of the process. Chip removal and transportation away from the machine can be accomplished with a defined auger system. Long, stringy chips — often an undesirable byproduct of machining ferrous metals — must be avoided through proper tool design, for they can be the source of possible heat distortion of the machine. If excessive rises in temperature cannot be avoided, it is prudent to provide temperature sensors that monitor the temperature at critical spots inside the machining area. Possible shifts within a few micrometers can then be compensated for by making cutting tool and/or part-fixturing adjustments. Dry-machining also generates dust particles. Filtration systems can keep dust to a minimum; but when dust becomes airborne, it has to be sucked out of the machine area. This area is, as a whole, separated from the spindle drive system and other sensitive hydraulic and electronic support systems of the machine, and by pressurizing this area slightly, dust can be prevented from entering (see Figure 2.7). Since some of the design features will be incorporated into any modern machine tool, the extra design cost compares favorably, indeed, to the benefit of proper machine design right from the start.

Applied Technology

The practical applications of different machining processes described below demonstrate how far dry-machining can be taken, how it can be

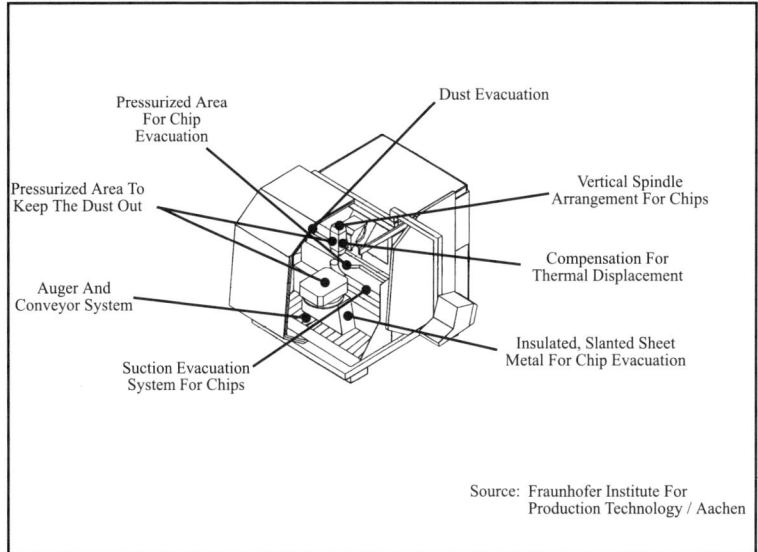

Figure 2.7 • Machine Design Features for Dry-Machining

approached, and how it can be done with relative ease. The right combination of the described technical variables guarantees success.

Turning: Among the main machining operations, turning is most widely used for dry cutting in a continuous, uninterrupted mode. Gray iron can be machined dry as well as wet, yielding equal finishes and tool life. Almost the same holds true for machining aluminum, steel, and hard cast iron without coolant as a substitute for grinding (known as "hard turning"). However, due to the thermal stress, the tool life of the cutting materials most often used — i.e., ceramics or PCBN (polycrystalline cubic boron nitride) — is somewhat limited and could stand improvement.

The challenge of machining extremely hard materials on regular production runs has led to the development of an intriguing new method incorporating a laser beam as an add-on package to existing machines, as illustrated in Figure 2.8.

Silicon nitride (Si_3N_4), with its extremely high tensile strength, makes any cutting material break down quickly due to thermal stress. However, by using laser-supported cutting, the cutting area can be prewarmed, and the resulting increase in temperature transforms the physical-mechani-

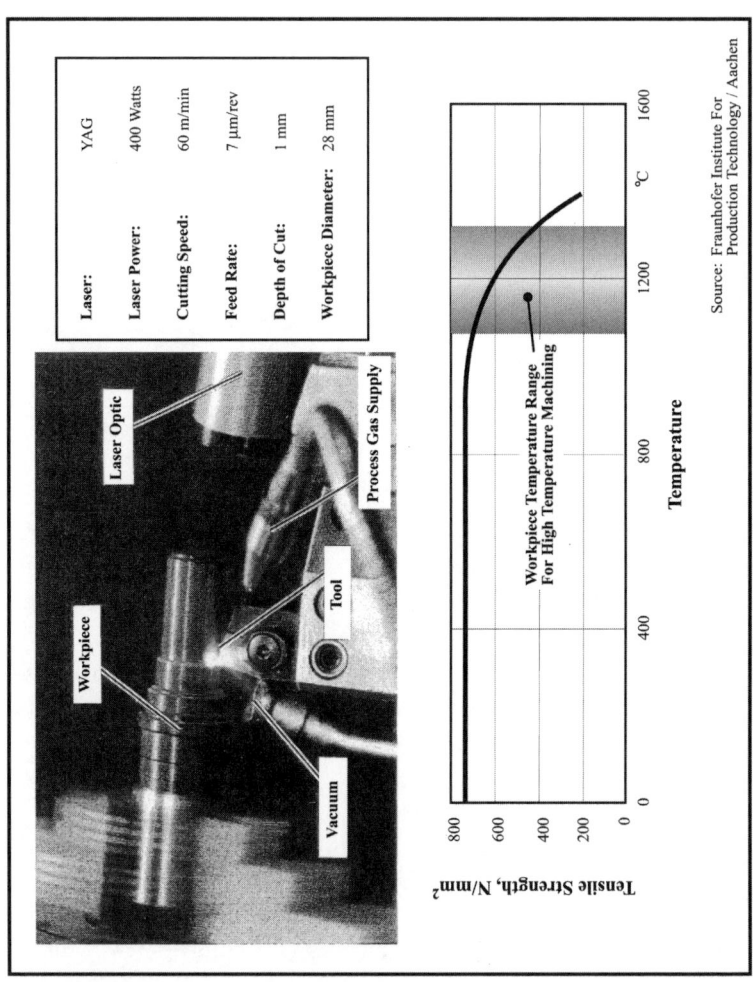

Figure 2.8 • Laser-Assisted Turning

cal characteristics of the workpiece material by plastifying it — decreasing its tensile strength and thereby improving the machinability of the material. Laser-assisted machining is a hot-machining process (between 800° and 3,000°C) that not only lowers the cutting forces but also reduces tool wear, decreases vibration, and allows for higher removal rates.

Using laser-assisted machining, turning operations are characterized by a 50% reduction in cutting forces due to increased temperatures in the chip cross-section achieved by softening the workpiece material. The wear on PCD and CBN cutting material when turning ceramics is somewhat higher because of the added workpiece temperature induced by the laser, but the same temperature rise causes the workpiece material to relax due to lowering the compressed residual stress through annealing and, therefore, achieving surface finishes comparable to those of grinding operations (approximately Ra=0.4 mm).

Generally, machinability can be improved with the aid of laser treating. Tests of milling ceramics with laser assistance reveal that cutting forces can be reduced by up to 70%, while tool wear decreases by about 80%. Milling of steel results in cutting force reductions between 30% and 70%, and substantial tool life increases when compared to conventional milling operations.

Titanium-aluminum-vanadium (Ti-6Al-4V) and reaction-bonded silicon-nitride (RBSN) are representative of extremely difficult-to machine workpieces. Both are sought after in the aerospace industry due to their hardness, high strength, low thermal expansion, and heat treatability. However, they both exhibit low thermal conductivity of 15 and 13 W/m^2 °C respectively. The high temperature generated during turning causes the cutting material to break down, leading to premature tool wear. Low metal-removal rates and the cost of the tools have led to a new method of machining developed by a research team at the University of Nebraska-Lincoln. This system, illustrated in Figure 2.9, uses a cryogenic cooling system that induces liquid nitrogen (LN) near the rear end of the cutting insert through a cap mounted on the tool holder that forms a chamber through which the LN flows in and out constantly.

Turning RBSN with PCBN and cryogenic "coolant" lowers tool wear substantially and improves surface finishes, since the cutting edge doesn't break down as quickly during the first cuts, as can be seen in Figure 2.10. Dry-machining of Ti-6Al-4V with a regular carbide insert and liq-

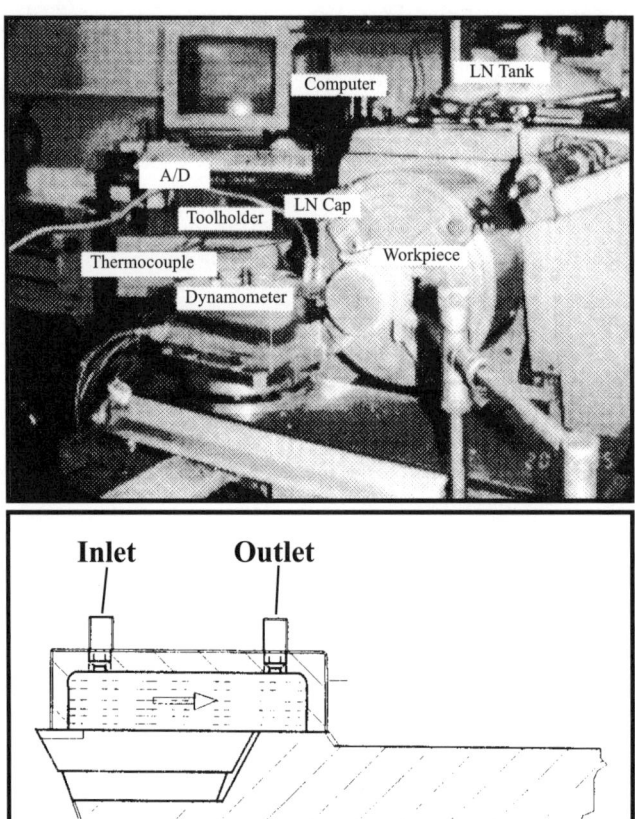

Figure 2.9 • Principle of Cryogenically Cooled Turning

uid nitrogen shows that the insert's flank wear occurs at a much slower pace.

The upshot is that the machining of difficult materials becomes more manageable and more productive when it is machined dry with the aid of an LN agent. The cooling effect reduces tool wear and prevents softening of the cutting insert.

Reaming/Fineboring: Reaming and fineboring are usually finish operations, with geometric part finishes being the most important

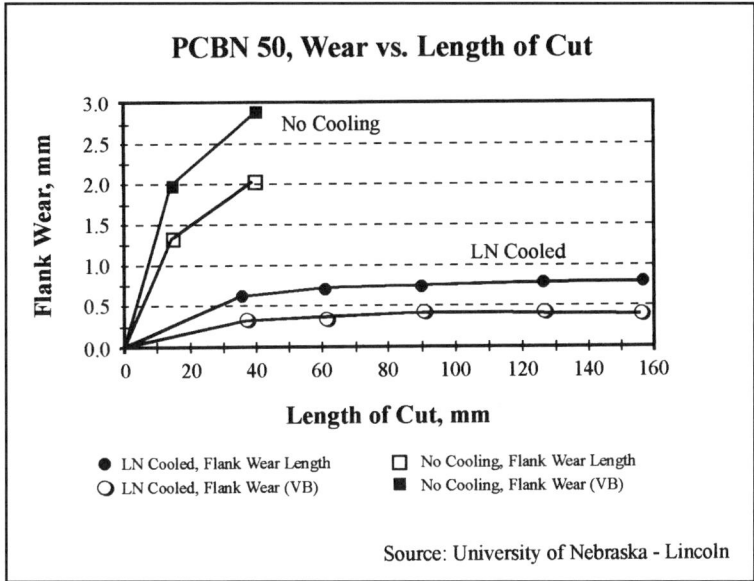

Figure 2.10 • Wear of PCBN Tool Turning RBSN

criteria. For close-tolerance machining, these tools feature indexable inserts and peripherally arranged guide pads. While dry-machining aluminum results in relatively poor surface finishes due to the material's adhesion to the insert and pad, dry-machining steel and cast iron can, however, achieve good results with proper tool design.

An example is the machining of the distributor and oil-pump bore of a 4-liter cast iron engine cylinder block, shown in Figure 2.11. Running on a transfer line, the tool design incorporates the HSK-connection with adaptation for radial and angular alignment. The guide pads ensure chatter-free machining and perfect alignment of the two bores, and the heavy metal tool bar minimizes the inherent tool vibration during cutting.

The keys for dry-machining the block are the oversized chip flutes that ease the chips out of the bores and above all the PCD guide pads — their chemical nonaffinity to cast iron prevents chip buildup, and their hardness minimizes the friction with cast iron and, thus, heat buildup. To stabilize the cut through the guide pads, a heavy metal bar is used in combination with minimal tool runout which results in successful finish-machining with CVD TiC-TiN coated carbide inserts. The tool life of up to 180 pieces per edge, and 360 per insert, covers one production shift.

58 • Chapter 2

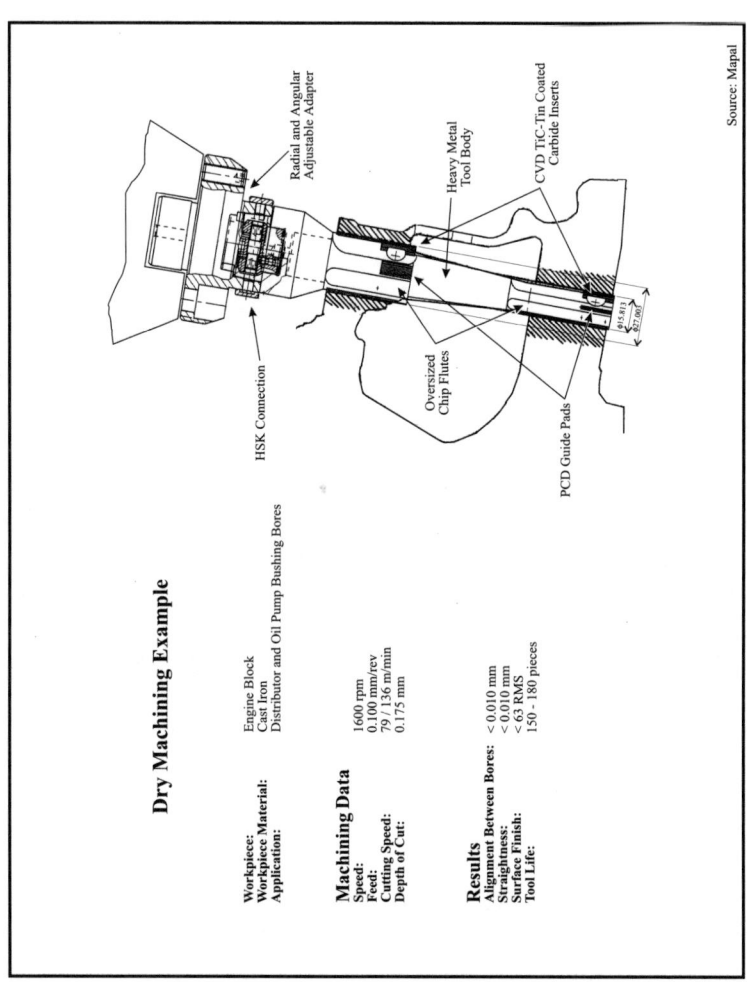

Figure 2.11 • Dry-machining with Fineboring Tool

Drilling: Dry-drilling (into the solid) is most problematic due to the difficulty of dissipating heat during the process and by chip transportation via helical flutes. Dry-drilling cast iron is manageable and yields tool life comparable to wet-machining; whereas with malleable cast iron, tool life reaches only about 75% of wet life expectancy. (See Figure 2.12.) Any reliable drilling in aluminum needs to be supported by minimal coolant supply due to aluminum's high degree of adhesion. When dry-drilling steel, the right combination of cutting parameters, cutting material, enlarged chip flutes, and induced pressurized air can make a difference, as demonstrated by numerous tests at leading technical universities which reveal the following results:

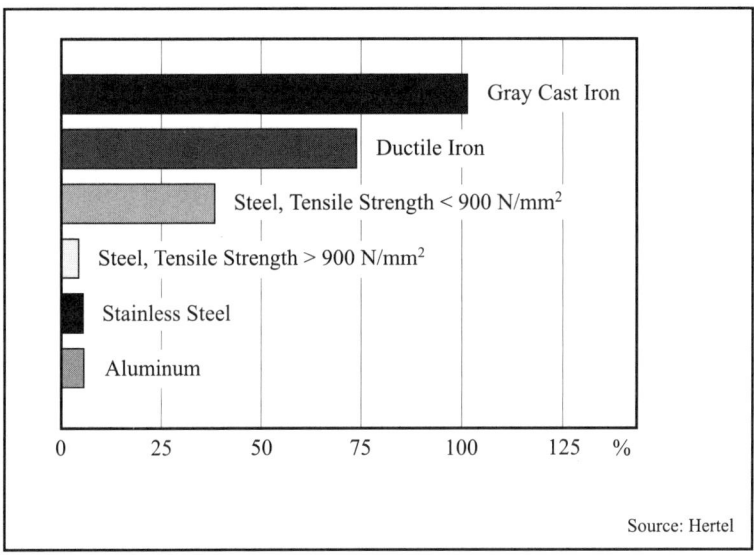

Figure 2.12 • Tool Life Changes with Dry-Drilling Compared to Drilling with Coolant

Parameters of low carbon steel workpiece material:
Bore diameter   8.5 mm
Cutting material   Cermet
Blind bore depth   25 mm
Cutting speed   90 m/min.
Feed rate   0.2 mm/rev.

Results: The accumulated bore length of the individually machined bore was 13.5 mm, which did not indicate the actual tool life. The cermet inserts at that point showed just slightly rounded cutting edges. The test was to reveal the influence of temperature on bore geometry and surface finish.

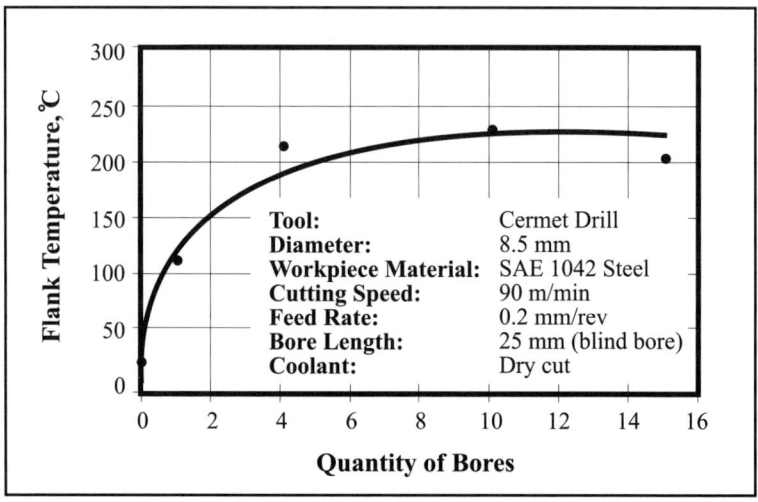

Figure 2.13 • Cutting Tool Temperature as a Function of Number of Bores Drilled

Temperature rise of the cutting material: After machining eight bores, the temperature on the material stabilized at about 225°C. At that point, a good part of the heat is absorbed by the cutting chips. The cutting force increased from 1.7 kN at the beginning of the cut to 2.2 kN at the end, while the torque increased from 8 Nm to 12 Nm. (See Figure 2.13.)

Bore Quality: Because of the heat generated, the cutting tool expands and cuts bigger bores with increased bore length. As shown in Figure 2.14, the maximum difference was 12 mm, which lies within the tolerance band IT7. A back taper slightly higher than normal prevents the drill from getting stuck in the bore when retracting. A comparison of the average surface roughness R_z at the beginning of the individual bores to the end of them shows a linear difference of 12 mm to 17 mm. The bore roundness was within 8.3 mm. The overall bore quality demonstrates

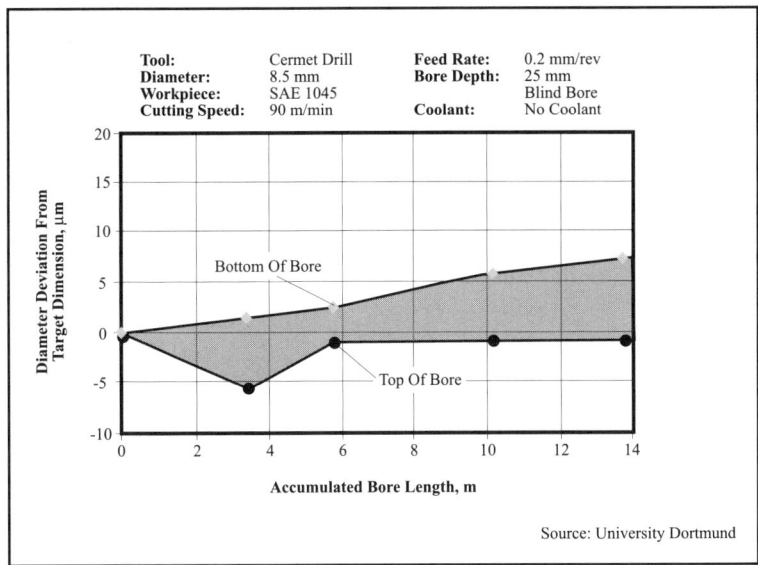

Figure 2.14 • Diameter Change Over Bore Length

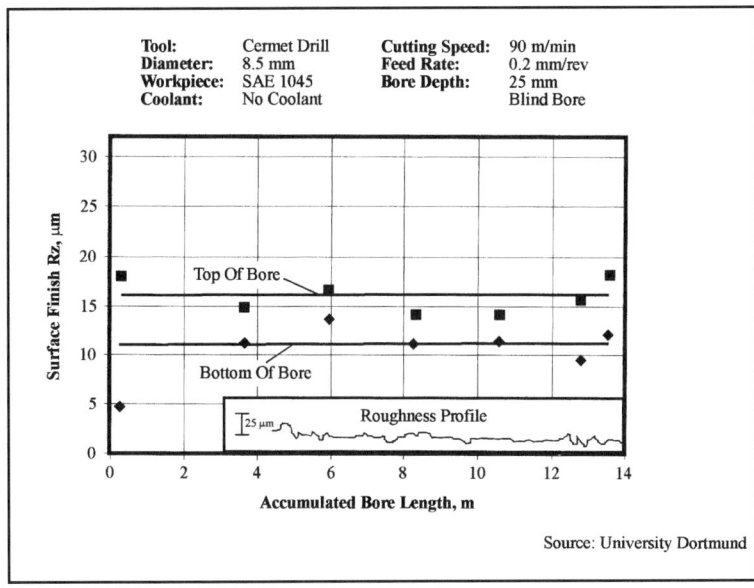

Figure 2.15 • Surface Finish Over Bore Length

62 • Chapter 2

that dry-drilling in steel can be done with relative precision (see Figure 2.15).

Frazzle-free dry drilling without delaminating is the yardstick for drilling composite materials. This can be accomplished with high-precision finished PCD cutting edges (see Figure 2.16) using the right cutting angle and fine-tuned machining parameters.

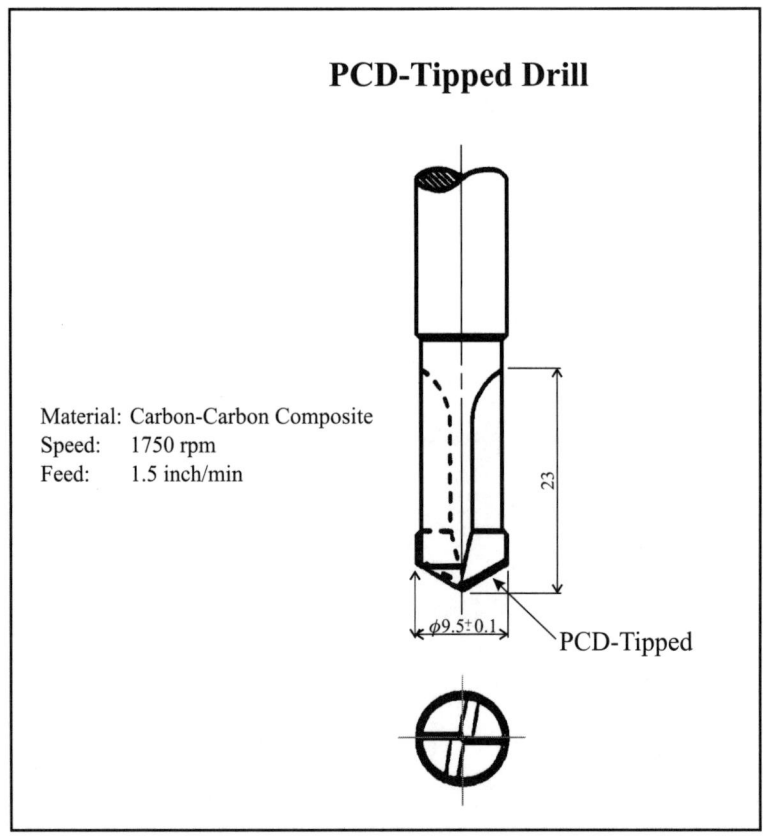

Figure 2.16 • PCD-Tipped Drilling in Carbon Composites

Milling: Milling is an operation of interrupted cuts, putting intermittent temperature stress on the cutters. The induction of coolant enhances that effect, which is why dry-milling usually yields longer tool life than wet-milling. This holds true for almost all popular workpiece materials except aluminum. As mentioned earlier, aluminum will adhere and weld to the cutting tool. For other materials, CVD and DLC coatings in conjunction with minimal coolant consumption are promising. The objective of dry-milling is to extend tool life as much as possible; and, for this application, silicon-nitride, carbide, cermet, and borazon are the cutting materials of choice. Further improvements can be achieved through suitable coatings, especially TiAlN and TiCN and the so-called "soft" coatings. Figure 2.17 illustrates the increase in tool life — machining with a six-fluted milling tool uncoated versus TiCN and soft-coating. Milling 70 m, or a milling time of four hours, will require the use of three soft-coated solid carbide tools instead of five "hard"-coated. Soft coatings can also offer higher cutting speeds and increased surface finishes — true determinants of productivity progress in finish-machining.

"Dry and High": The key to combining economics with productivity could be machining without coolants at high cutting speeds. At high cutting speeds, the wear on the cutting material is even more determined by its adhesion to the workpiece material and the cutting temperature. The finish geometries — tolerances and surface quality — are dependent on the temperature induced into the workpiece material. One-pass machining with light cuts and high feed rates, together with high cutting speeds, is preferred. Milling and turning operations can be applied, especially in cast iron. The cutting materials silicon-nitride and cubic boron nitride are best suited. The latter has a higher purchase price, but can be used for longer cuts.

(NEAR)-DRY-MACHINING

When dry-machining is not technologically feasible or is economically imprudent — particularly for drilling, reaming, and fineboring — then near-dry-machining is an interesting alternative. The idea is to induce a metered, minimal amount of lubricant onto the cutting edge. This principle of "minimum volume lubrication" (MVL) replaces conventional flooding during machining. It is, in fact, a near-dry process.

Typically, machining centers handle coolant volumes of 20 to 100 l/min. With MVL, only 0.03 l/h to a maximum 0.2 l/h of fluid is used. The lubricants are vegetable-oil based, nontoxic, and biodegradable. Their application is safe, controllable, and precise, and can be monitored with devices that automatically check temperature and dimensional part de-

64 • Chapter 2

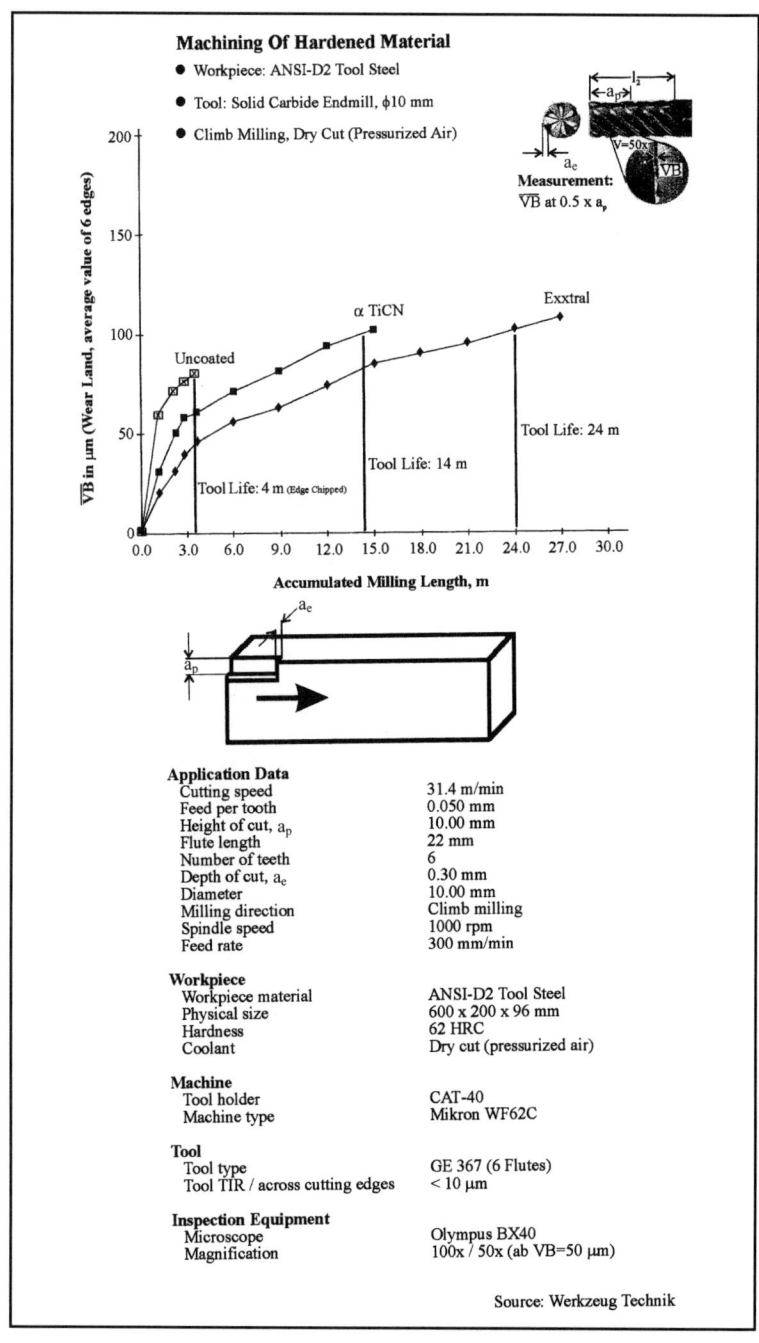

Figure 2.17 • Machining Data/Machining Results of Milling with "Soft" Coating

viation from the blueprint. There are no expenses for collecting, treating, or disposal of coolant, and chips produced by this method are considered dry, so no reclamation costs are incurred.

In near-dry-machining, small lubrication particles are mixed with air and are directed to the cutting tool from either outside the machine or through the machine spindle. The advantage of directing the lubricant from the outside is that no major modifications of the machine are necessary. The disadvantage is in the length to diameter ratio of about 2 to 2.5, which means that no lubrication can be brought to the cutting edge when machining bores. For through-the-spindle lubrication, it is necessary to design a spindle for dry rotation, but this method offers a better controlled and versatile process. See Figure 2.18.

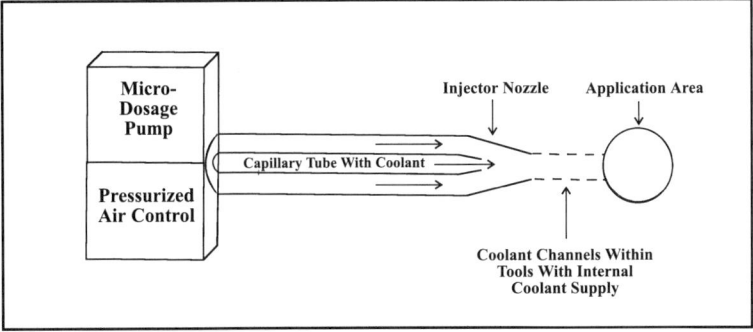

Figure 2.18 • Application of Minimum Volume Lubrication Supply

In MVL, cooling plays a smaller role than lubrication because of the hot hardness of advanced cutting materials, and because temperatures transferred to the workpiece can be kept to a minimum with the right cutting parameters. MVL mostly lubricates the process and, by reducing the friction, it curbs the cutting temperature and lowers the wear on the tool. The lube film separates the tool from the workpiece and effectively prevents adhesion.

As shown in Figure 2.19, the oil/air mix can be induced into the tool on the cutting edge principally in four ways.
1. The oil/air supply is brought onto the cutting edge from the side. Although simple, this system is unreliable since the air exits outside the impact area and there is no support for chip discharge.

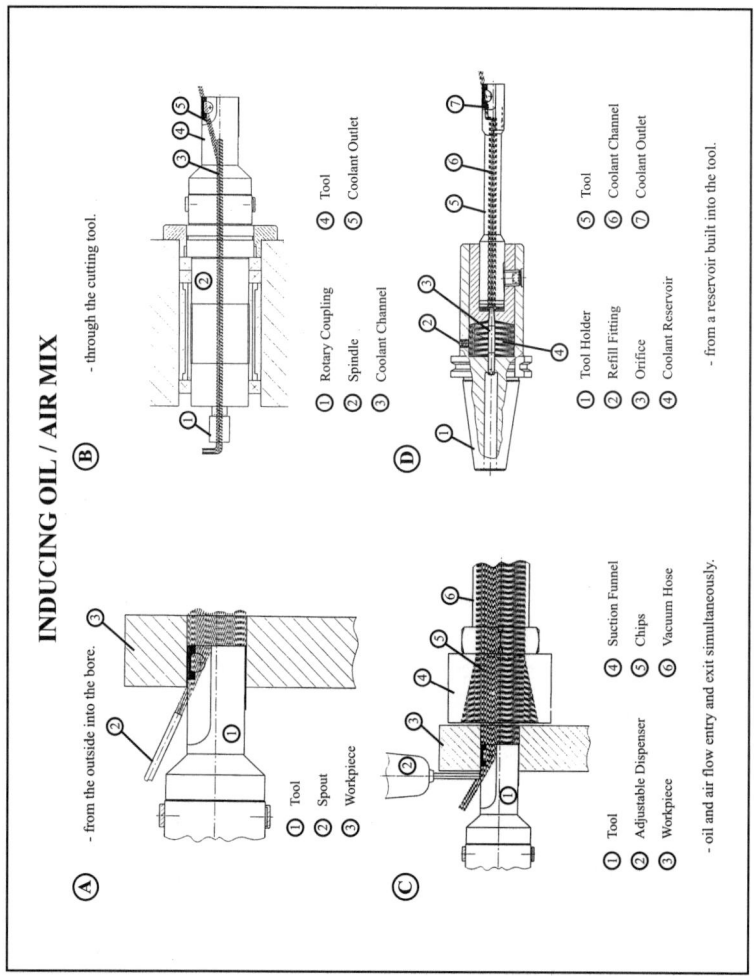

Figure 2.19 • Various Methods of Inducing Oil/Air Mix

2. The advantage of lubricating through the cutting tool is that only a few parts are required to induce the oil/air mix. The disadvantage is that for long tools, the lubricant has to travel a long distance and could get stuck in the supply pipe. To prevent this, the oil concentration will have to be increased, and that could lead to undesirable mist formation.
3. Using a vacuum system is the most versatile method because of its simplicity and reliability. A suction hose at the workpiece guar-

antees clean and safe chip disposal, while the oil and air mix is precisely metered to a fine spray. For flexible manufacturing involving many different parts, the chip suction feature might be a bit too complicated.

4. When using a reservoir within the tool, the lubricant becomes an integral part of the tool holder and it sucks particles out through the airstream.

Applied Technology

Near-dry-machining with minimum volume lubrication establishes a new direction in machining technology. Practical applications in drilling and reaming (fineboring), and the challenge to finish-machine aluminum, illustrate that this process has great potential.

Drilling

Comparing cutting force measurements (see Figure 2.20) of dry versus regular coolant versus MVL for drilling in steel shows that when using the MVL process, 8 ml/h oil/air guarantees enough lubrication for the entire drilling operation. It is interesting to note the dynamics of the

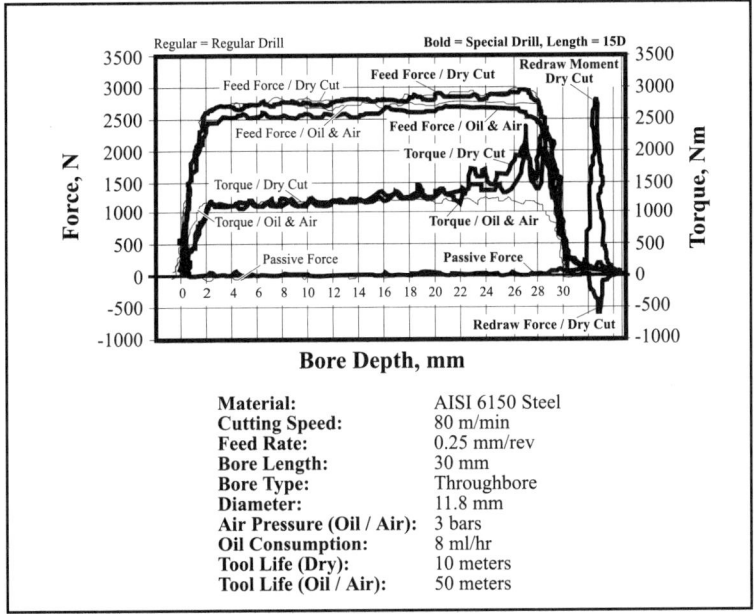

Figure 2.20 • Cutting Forces Dry-Drilling versus Drilling with MVL

diagram of dry-drilling as the bore penetration gets deeper. The tool life is equal to that of regular coolant-fed drilling (50 m bore drilling length) and 5 times as high as drilling dry.

As shown in Figure 2.21, the wear on the tool is the same for machining with regular coolant and with MVL. In fact, using wider spiral flutes for easier chip transportation in conjunction with TiAlN tool coating and minimum volume coolant constitutes significant technological progress since the surface finishes do not vary between the two different processes.

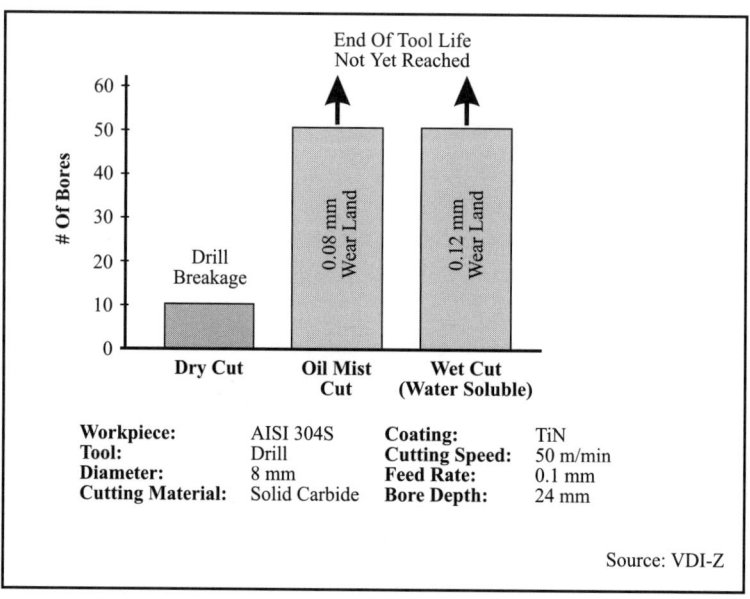

Figure 2.21 • Tool Life Drilling Aluminum with MVL versus Regular Coolant versus Dry

Reaming

Precision finish-machining of critical tightly toleranced bores on regular production machines is often done with reaming tools featuring peripherally arranged guide pads. To guarantee sufficient lubrication, adequate lubrication holes and exits for pads and inserts have to be designed for the small amount of MVL available. Pad material should be PCD for ferrous and nonferrous material, and the inserts should be coated carbide for cast iron, cermet for steel, and PCD for aluminum.

The function of chip transportation in MVL is rather limited because of its lack of volume and density. The right insert geometry (chipbreakers) together with the right machining data will ensure the formation of small chips. When reaming as a fine-finishing operation, the data and geometry should be rather conservative, thereby limiting the heat generated during cutting. The emphasis is on heat prevention rather than heat discharge or transfer. Compared to conventional coolant usage, practical applications have shown that reaming with MVL can be done with almost identical machining data using inserts coated with TiN, TiAlN, TiC, and the right insert geometry.

Varying conventional coolant supply and MVL, while leaving all other machining parameters constant, causes no significant differences in bore dimensions and surface quality when machining steel and aluminum. See Figure 2.22.

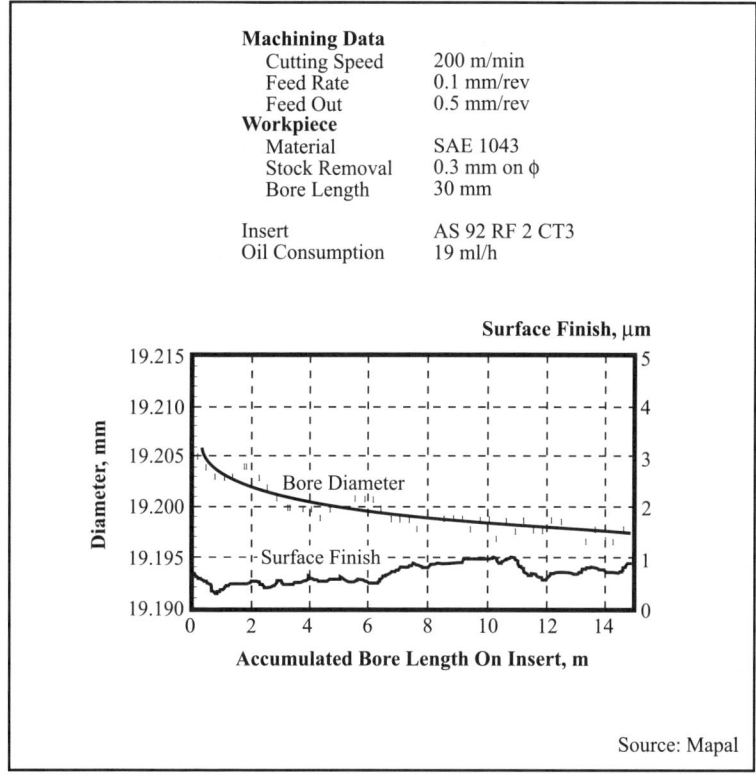

Figure 2.22 • Reaming in Steel with MVL

Empirical tests of fineboring with padded, indexable insert cutting tools in different workpiece materials are indicators of sensible MVL usage (see Figure 2.23). Those same tests also reveal that fineboring tools with inner coolant passages, using conventional coolant, can change the original grade during operation, and that fineboring tools with complex inner coolant passages using conventional coolant supply can change the original balance grade at high speeds due to the coolant's mass. In these cases, MVL is more suitable. However, at high speeds, reliable oil/air mixes need a minimum of 6 bar air pressure.

How Does MVL Fare in Finish-Machining Aluminum Bores?

A comparative study, shown in Figure 2.24, of different coolant supply principles used to finish-machining bores in aluminum ($AlSi_9Cu_3$) through reaming reveals interesting facts. The study investigated five alternatives:

1. machining dry,
2. MVL induced from the outside (MVLO),
3. MVL induced from the outside plus pressurized air through the tool (MVLOA),
4. MVL induced through the tool (MVLI), and
5. regular emulsion through the tool.

Measurements were taken of bore tolerance and surface finish of the workpiece. Reamers with PCD blades and carbide guide pads show substantial diameter variations and rough surface finishes when machining dry. This can be improved by using PCD guide pad material to partially forego aluminum pickup through adhesion. However, neither indexable PCD blades nor PCD-tipped multiflute reamers are sensibly workable dry. A noticeable improvement can be achieved through the application of minimum coolant lubrication from the outside with pressurized air through the tool up to a length-to-diameter ratio of 2:1. Beyond that, a lack of proper lubrication deteriorates the surface finish.

In addition, outside-MVL supply requires up to 10 times the amount of oil/air compared to through the tool supply, and directing the flow parallel and to the tip of the tool is also difficult with outside supply. Supply through the tool is both reliable and efficient. Applying regular emulsion shows no improvement over well regulated and metered MVL for both bore diameters and surface finish. In fact, the process temperatures using emulsion and MVL are nearly even, and both

Examples of Precision Fineboring With "Minimum Volume Lubrication"

	Gray cast iron ASTM / A48, CL 40	Pure iron	Pure aluminum	Stainless steel AISI / 422	Steel ASTM / 1060	Gray cast iron
Bore, Dia. [mm]	130 Dia. x 130	35 Dia. x 60	20 Dia. x 60	20 Dia. x 50	20 Dia. x 50	20 Dia. x 50
Tolerance	K6	P7	H6	-	-	-
Target [mm]	Mean	-	20.006 - 20.009	-	-	-
Guide pad material	PCD	Cermet	PCD	PCD - Cermet	PCD - Cermet	PCD - Cermet
Insert material	Cermet	Cermet	PCD	PCD - Cermet	PCD - Cermet	PCD - Cermet
Cutting speed [m/min]	400	250	800	200 - 220	200	125 - 175
Feed / Rev. [mm]	0.08 - 0.1	0.1	0.1	0.05 - 0.125	0.025 - 0.1	0.05 - 0.1
Depth of cut [mm]	0.3 - 0.5	0.15	0.1	0.1 - 0.15	0.1 - 0.15	0.1
Insert geometry - Radial Rake Angle - Lead Angle(s)	Hexagonal	12° 3°/30°	6° 75°	8° 3°/30°	12° 3°/30°	6° 3°/30°
Surface Finish, Rz [μm]	2 - 5	< 5	< 2	< 1	< 1.5	< 1.5
Bore geometry [μm] - Circularity - Cylindricity	4 - 7 / < 4 / < 6	< 4 / < 3 / < 2	< 1.5 / < 2 / < 2	< 2 / - / < 3	< 2 / - / < 3	< 2 / - / < 3
Coolant Usage	< 1 ml/hour	-	1 ml/hour	< 0.3 ml/bore	< 0.3 ml/bore	< 0.3 ml/bore
Coolant introduced	external	central	central	central	central	central
Air pressure [bar]	6	6	6	6	6	6

Figure 2.23 • Precision Fineboring with MVL

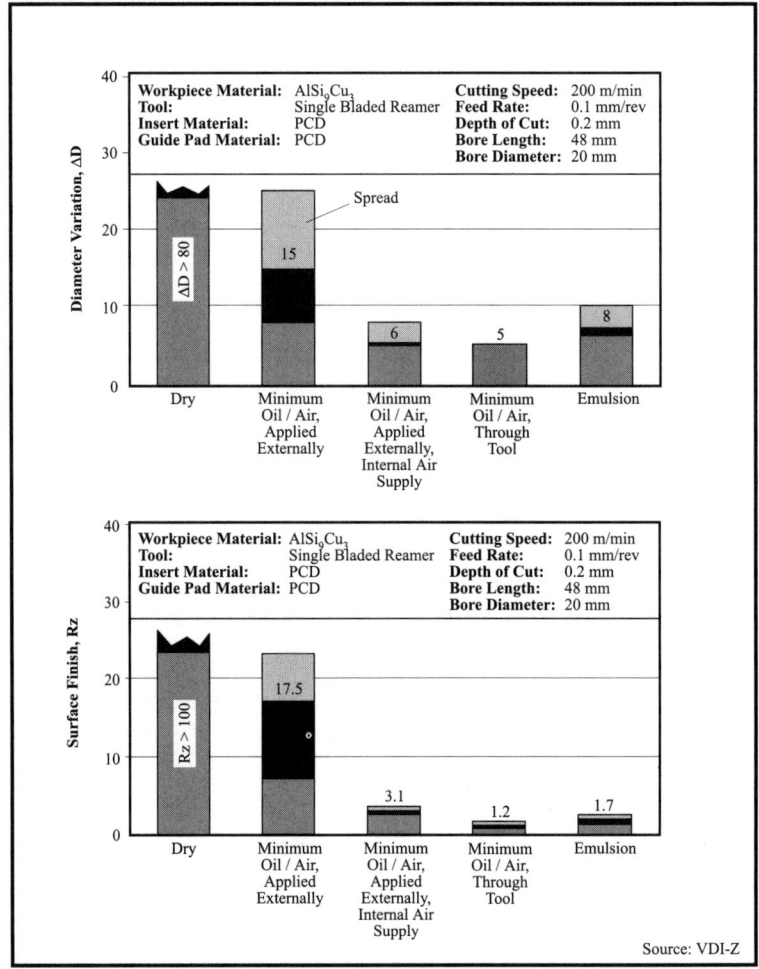

Figure 2.24 • Geometric Finishes

have a comparable influence on the structure of the workpiece material. MVL is clearly a reliable substitute for conventional use of emulsion in the machining of aluminum bores.

SUMMARY

Reductions in the amount of metalworking fluid in use is an important step toward more efficiency and productivity on the production floor. Advancements in machining technology have opened new avenues that manufacturing needs to pursue. Because retrofitting machines for dry- or near-dry-machining is relatively inexpensive, new machines should incorporate the design features for MVL, and cutting tool technology must be fine-tuned to guarantee its success.

Chapter 3

PRECISION ONE-PASS MACHINING

In the early days of modern manufacturing, machines were built to accommodate heavy cuts for roughing, semi-finishing, and finishing operations. The overall sizes of workpieces were larger, and workpiece materials were mostly ferrous metals and their alloys. The machining paradigm has changed, and today's foundries turn out unfinished workpieces much closer to finish dimensions. The trend has been for machine tools to become increasingly smaller for lighter cuts, while workpieces have also become smaller and lighter by design and through the use of more nonferrous materials. Machine spindles turn much faster, and machine axes have acceleration rates measured in g's. But manufacturing productivity is not measured in rpm's and g's — it is measured by how fast a finished part can be produced. Furthermore, the overall performance of manufacturing is measured by output per hour, and at what quality level and cost it is achieved. Consequently, output, quality, and cost of the finished product are a function of applied machining processes.

A radical departure from traditional machining is needed to produce products of (predictable) high quality using low cost finish-machining within minimum time frames. The principles of agile and lean manufacturing invite rapid machining and minimum capital expenditures. Today's machine tools — and those of the future — will predominantly be arrangements of machining centers with small to medium output ranges of 10 - 20 kW main spindle motors, running at medium to high speeds, and table (pallet) sizes of 500 - 800 mm. The "plug and play" approach, which groups a number of standard, stand-alone CNC machines, speeds up the inception of a new manufacturing system that will result in rapid, continuous production.

Frequent tool changes can offset gains in productivity related to high accelerations and speeds. In fact, noncutting times have to be kept to an absolute minimum to justify equipment for reduced main machining time. For reasons of logistics (handling, inventory, management), the number of cutting tools also has to be kept to a minimum. Furthermore — and contrary to common applied practice — the minimum number of tools available for a particular process can assure the best and easiest achievable part quality.

One-pass machining has to be pursued for any process, whether the tool is to produce a finished part from a raw casting in one pass, or one tool is to finish several configurations in one pass (combination tooling), or if one tool is used to perform several functions (e.g., circular milling).

FROM TRADITIONAL TO ADVANCED MACHINING

In the last several decades, there have been dramatic shifts in usage of

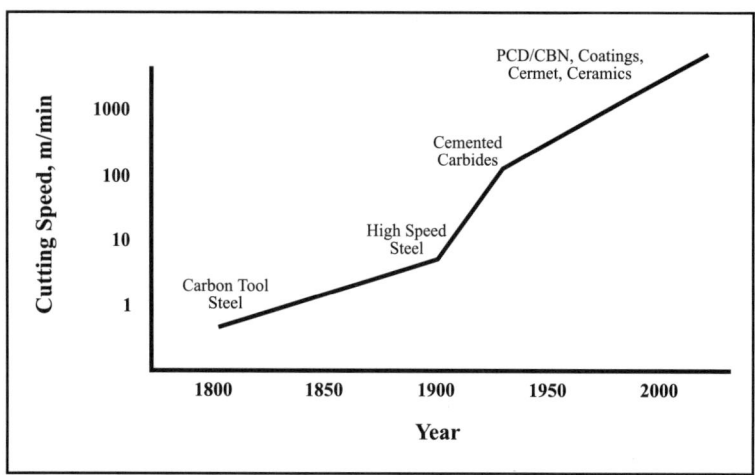

Figure 3.1 • Increases of Cutting Speed

workpiece materials, cutting tool materials, and applied machining data, and in the accuracy of machine tools.

The consumption of nonferrous metals, notably aluminum and magnesium alloys, has increased at the expense of ferrous metals. Engineered materials such as composites and so-called superalloys (usually titanium based) have seen increased usage. In order to get parts off the machine and have them completed within shorter time frames, cutting speeds have increased exponentially over the past decades (see Figure 3.1). This has been achieved thanks to the machine's improved accuracy and the development of harder, more wear-resistant cutting materials.

In the early days of mass-produced products designed for subsequent assembly line setups, parts and components had to be dimensionally defined and toleranced to make them fit in random order. Then, for automatic assembly, those parts had to be produced to even tighter tolerances. The requirement of producing parts, components, and subassemblies with predictable results — and without debugging them — is due to the ever-pressing need for lowering costs and staying competitive. Today, the only sensible approach is to produce high quality parts that are "right the first time."

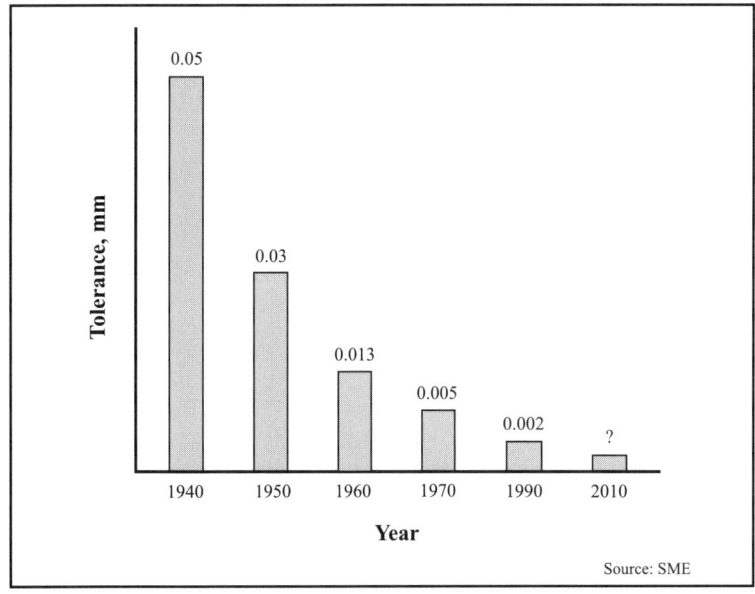

Figure 3.2 • The Narrowing of Tolerances

78 • Chapter 3

In order to meet demands for
- improved safety,
- extended service life,
- weight and noise reductions,
- emission control, and
- energy savings,

dimensional tolerancing is an important quality issue. (See Figure 3.2.)

For most applications, the old-fashioned method of specifying different operations for a sequential mode of machining can easily be abandoned, and finishing parts from the raw, unfinished cast can, in many cases, be done with one single pass. Advances in machining technology not only make this possible, they even invite it.

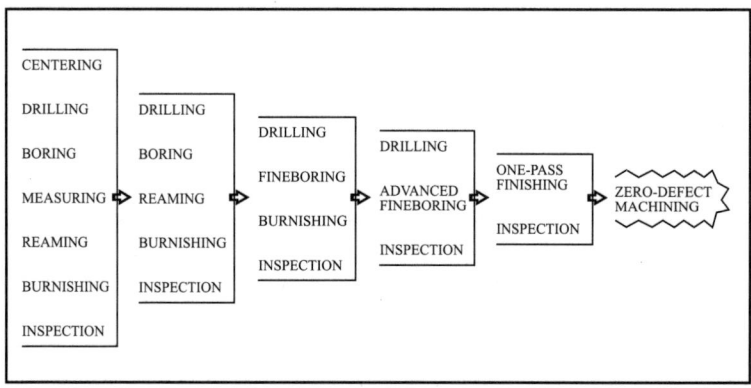

Figure 3.3 • The Evolution of Machining

Advanced Machining Technology

The following is a description of the different areas which comprise technology essential for successful one-pass machining.

Cutting Material and Workpiece Material

See Figure 3.4 for a summary of the advanced cutting materials covered below.

Carbides: The most popular cutting material — with 19 different grades — carbides cover a wide range of part material and machining parameters. Depending on the mix of cobalt, niobium, titanium, and tantalum, their hardness (higher grade numbers) and roughness (lower grade numbers) vary. Lower numbered grades are more suitable for heavier (roughing) cuts, while higher numbered grades are more wear resistant and are used for finishing cuts. So-called micrograin carbides are effective substitutes for regular carbide and can successfully be applied for extra high stock removal and high abrasive wear machining.

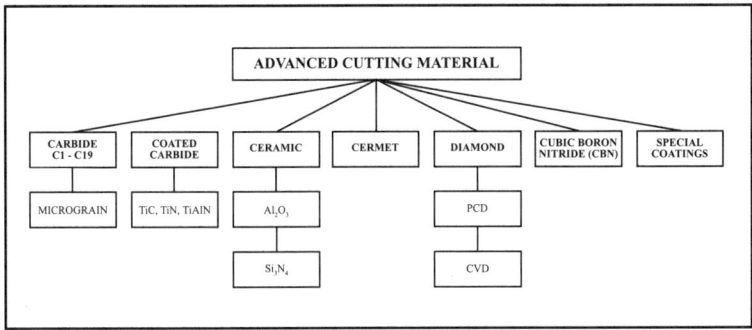

Figure 3.4 • The Array of Advanced Cutting Materials

Coated Carbides: Generally, TIC/TIN coated carbides offer tool life up to four times longer than uncoated carbides. Coatings also increase the cutting tool's versatility, since fewer grades are required to cover a broader range of machining applications. Depending on the applications, coatings can vary in thickness and layering; and multilayer coatings can increase heat, wear, and crater resistance, thus prolonging tool life.

Ceramics: Ceramics are available with two base materials — either aluminum oxide (Al_2O_3) or silicon nitride (Si_3N_4). Their brittleness limits their usage somewhat, but they are ideal for machining cast irons and some steels up to HRC60 with high cutting speeds. Whisker-reinforced aluminum oxide (Al_2O_3 + SiC) increases its otherwise low thermal stress coefficient and combines higher toughness with high wear resistance, making it well suited for tough nickel- and chromium-based steel alloys. Silicon nitride is excellent for machining gray cast iron. Its relatively high thermal factor and high toughness allow it to machine part interruptions even at higher feed rates.

Cermets: Carbides, containing either niobium, tantalum, or molybdenum, are added to a titanium nitride base to form cermet. Although their hardness is similar to that of carbides, cermets are less susceptible to diffusion and wear, and display better frictional behavior. Cermets are good for light to medium cutting in turning, milling, and finish-reaming applications at high cutting speeds.

Cubic Boron Nitride (CBN): High pressure, high temperature sintering of CBN particles and a binder material is used to form cubic boron nitride. In hardness, CBN ranks second only to polycrystalline diamond (PCD). Unlike PCD, it has a low affinity to ferrous metals and stays chemically stable, even when machining at high cutting speeds. Furthermore, CBN has high temperature resistance and heat conductivity, and it can prefinish and finish-machine hard cast iron (nodular and malleable) as well as superferrous alloys with high cobalt and nickel content. Parts which were once finished by grinding can now be more productively machined by turning, milling, and (fine)boring. Certain cast irons can even be dry-machined with CBN, thus foregoing costly coolant provisions.

Polycrystalline Diamond (PCD): Polycrystalline diamond consists of a layer of diamond powder, sintered to one uniform mass of 0.5 - 0.7 mm thickness which is pressed onto a substrate of cemented carbide. By using different grain sizes of diamonds, it is possible to optimize PCD for specific applications. The microstructure of the PCD product influences its physical characteristics, that is, the coarser the grain, the better the abrasion and impact resistance.

PCD tooling is used to machine nonferrous, nonmetallic material. It is extremely popular for machining light metals, especially aluminum. Even abrasive high-silicon content aluminum (12% - 20% Si) can be machined with ease and excellent results in terms of surface finish and tool life with the tool running at high speeds can be achieved. Diamond tooling requires a rigid setup of machine tool and part fixturing, not-too-high feed rates, and a moderate depth of cut. However, they should be run at very high speeds. This, combined with a good soluble cutting fluid, yields up to 300 - 400 times the tool life of carbide grades in finish-machining operations.

PCD-tipped cutting tools are applied with selective workpiece materials, due to their superior hardness and wear resistance as well as fracture strength for good dimensional part control and excellent workpiece finishes.

A number of special coatings, notably soft coatings (such as MoS_2) that further lubricate the process, have been developed and successfully used for special application, such as dry-machining. Other coatings have been developed for cermet, and even cubic boron nitride, in an effort to further increase tool life for certain applications and specific workpiece materials.

Optimized Cutter Geometry

Modern indexable inserts are "well-balanced" compositions of angles, radii, flats, and curvatures, reflecting careful optimization of fixed and variable machining conditions. This is especially useful in the area of finish-machining and one-pass machining, where the optimum cutting geometry can be adapted for special conditions. Sometimes, this means choosing from what is already available by combining cutter shapes, sizes, angles, etc., and applying the machining data that best suit the design, or vice versa. Often, when parts are to be finished with one operation, new designs have to be created and tuned to different machining parameters.

Figure 3.5 depicts a meticulously designed cutting insert conducive to one-pass machining. The primary and secondary lead angles provide the desired surface finish, and increased side rake angles lower the cutting force. Face clearance and relief angles should not be greater than 5° - 7° so as not to weaken the insert. The clamping angle prevents movement of the insert during cutting, while the chipbreaker aids in breaking other chips into manageable sizes.

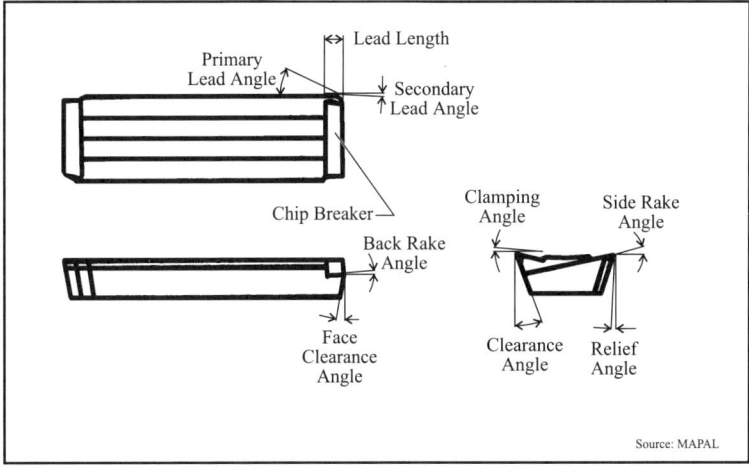

Figure 3.5 • Insert Geometry

Chip control is an important criterion for selecting the appropriate machining data. The process of chip making is basically that of deformation, and the plastic flow of material generates a smooth, continuous flow of chips along and past the tool face that carries with it a big part of the thermal stress. Breaking the chips at a point where neither size nor form can negatively impair the part surface allows the chips to be easily discharged. High feed rates create higher material distortion and deflection to aid in breaking and controlling chips, and higher cutting speeds entail increased cutting temperatures which soften the workpiece material. The result is that cutting can be done more easily, and a smooth flow of chips is produced.

High cutting speed invites one-pass machining because it can generate good surface finishes in a shorter amount of time.

Well-Defined and Balanced Tool Body

In one-pass machining, there is no provision for correcting machining passes. Consequently, the tool's design has to be "perfect." This includes the tool body itself. Clamping plates must be flush with the tool body because any protrusion can hamper chip flow and adversely affect the outcome of the process. Any mechanism to adjust cutting inserts has to be allowed for in the tool body as well. Chip flutes have to be spacious, and areas for "natural" chip flow must be provided — the tool's coolant passages have to be directed precisely onto the cutting edge for the desired cooling and lubricating effect, and to aid in flushing out the chips. In the event of "padded" tooling, grooves alongside the guide pads will provide a lube film around the entire tool body.

In a precision machining environment, machine tool and cutting tool need to be balanced according to their rotational speed, allowing only minute imbalances specified by a given balancing grade. What is regularly overlooked is the tool's weight and its inherent vibration during cutting. The majority of tool bodies are high speed steel or carbide in the case of some solid tooling, while inserted tooling is usually regular tool steel (4140). For finishing operations, when vibration or weight becomes a specific issue, alternative body material has to be considered.

Vibrations, inherently generated by the machining process, can be decreased by the use of "heavy metal" body bars. Heavy metal is a high density material consisting of 90% - 93% tungsten, with the rest composed of roughly equal amounts of iron and nickel. Its density is about

17 grams/cm³ with a hardness of HFC24 and a modulus of elasticity of 40 × 10⁶ psi (275 GPa). The equation for deflection mathematically proves that a higher modulus of elasticity positively influences the rate of deflection. Padded tool bars of heavy metal can feature an amazing length to diameter ratio of up to 12:1 and still run "chatter-free" even at speeds as high as 5,000 - 6,000 rpm.

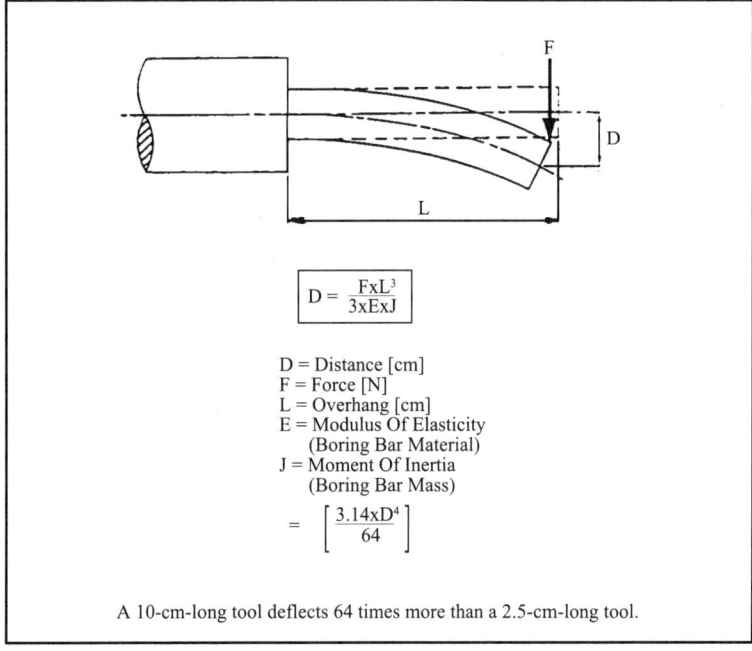

Figure 3.6 • Tool Deflection

Weight limitations are imposed on tooling systems (mostly in flexible manufacturing) when automatic system changes are required. The alternatives that should be considered are hollow bars and aluminum bars. Hollow bars can substantially reduce weight, but in a precision environment their use is sometimes limited since geometrically defined coolant passages and provisions for "combination" operations cannot always be accommodated. Aluminum bars, surface hardened with built-in steel cassettes or cartridges to incorporate the cutting inserts, make for light yet rigid tool assemblies.

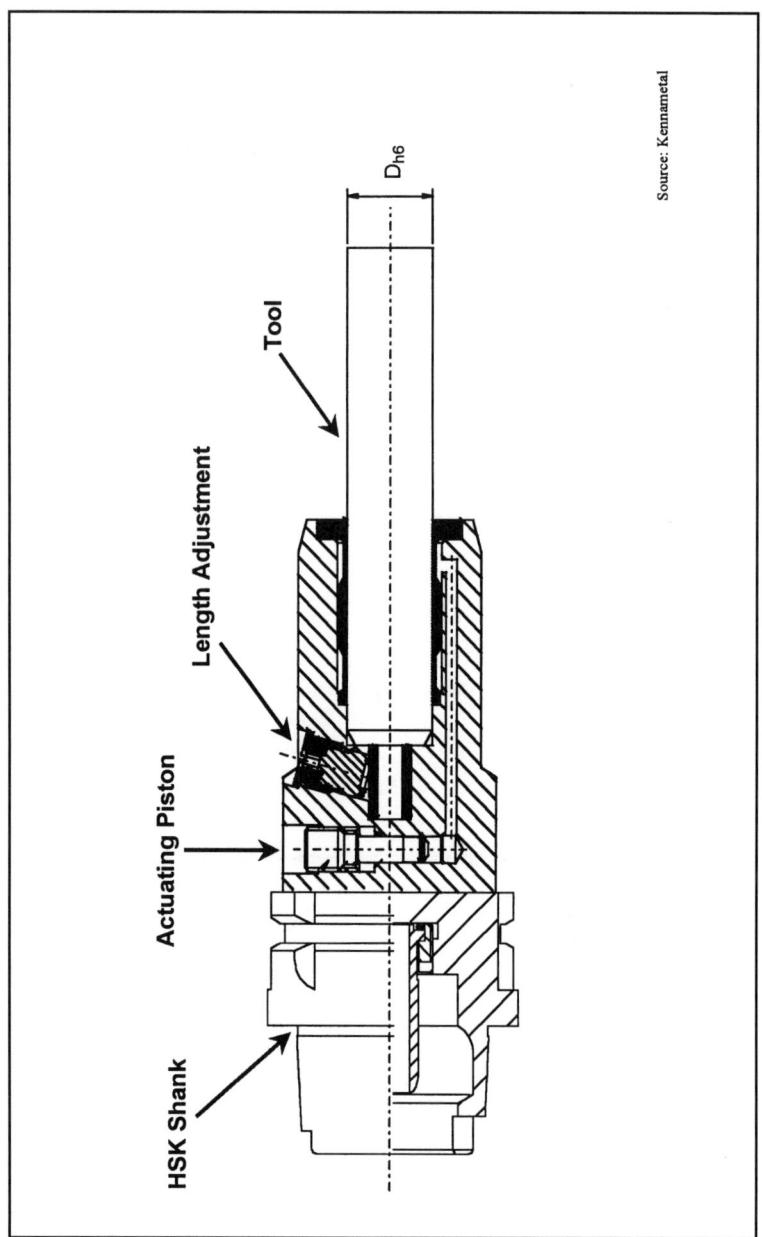

Figure 3.7 • Runout is Minimized with Tools Featuring High Static and Dynamic Stiffness

Accurate Toolholding and Tooling Interface

In flexible manufacturing, machining centers equipped with multiple tool pockets phase tools into the machine spindle within a matter of seconds, i.e. tool change times are between 1 and 2 seconds. The repeatability accuracy of precision tools must be within 2 - 4 µm with exact axial location and insert orientation. Shrink fit or hydraulic chucks are chosen for round-shank tools which require that the back part of the tools be designed to feature repeatable precision tooling interfaces (Figure 3.7).

When manufacturing with C_{pk}-2 or better, close blueprint tolerances are cut by half or more. This leaves no room for error, so tooling interfaces need to have high repeatability accuracy, minimum runout, and high static and dynamic stiffness allowing for reliable one-pass machining at high cutting speeds and increased depth of cut. The recently developed hollow taper shank HSK has set new standards along these lines (see Chapter 1).

As dedicated traditional transfer lines are replaced by flexible machining systems in modern manufacturing, cutting tools must have the ability to cut accurately without bushing support in a free-standing mode. Tools with extreme length to diameter ratios measure some degree of runout from the machine spindle face. To keep the radial or angular runout to a minimum, in-between adapters acting as bushing supports allow setting of the tools to "zero" before machining begins. An example of a high-precision tool assembly with in-between adapters and state-of-the-art interface for finish-machining multiple bores in one pass is shown in Figure 3.8.

Process Monitoring

There are specific devices built into advanced machining centers with the goal of "error free" machining to enhance one-pass precision machining. They work by monitoring the thermal growth of machine parts, gauging the workpiece data, and controlling feed rates. By monitoring the thermal growth of the machine tool's ball screw, positioning accuracy can be secured.

Preloaded ball screws generate considerable heat that can lead to length changes and other dimensional machine inaccuracies. Linear scales (encoders) mounted close to and alongside the ball screws can provide an absolute measure of the machine axis position and accurate part posi-

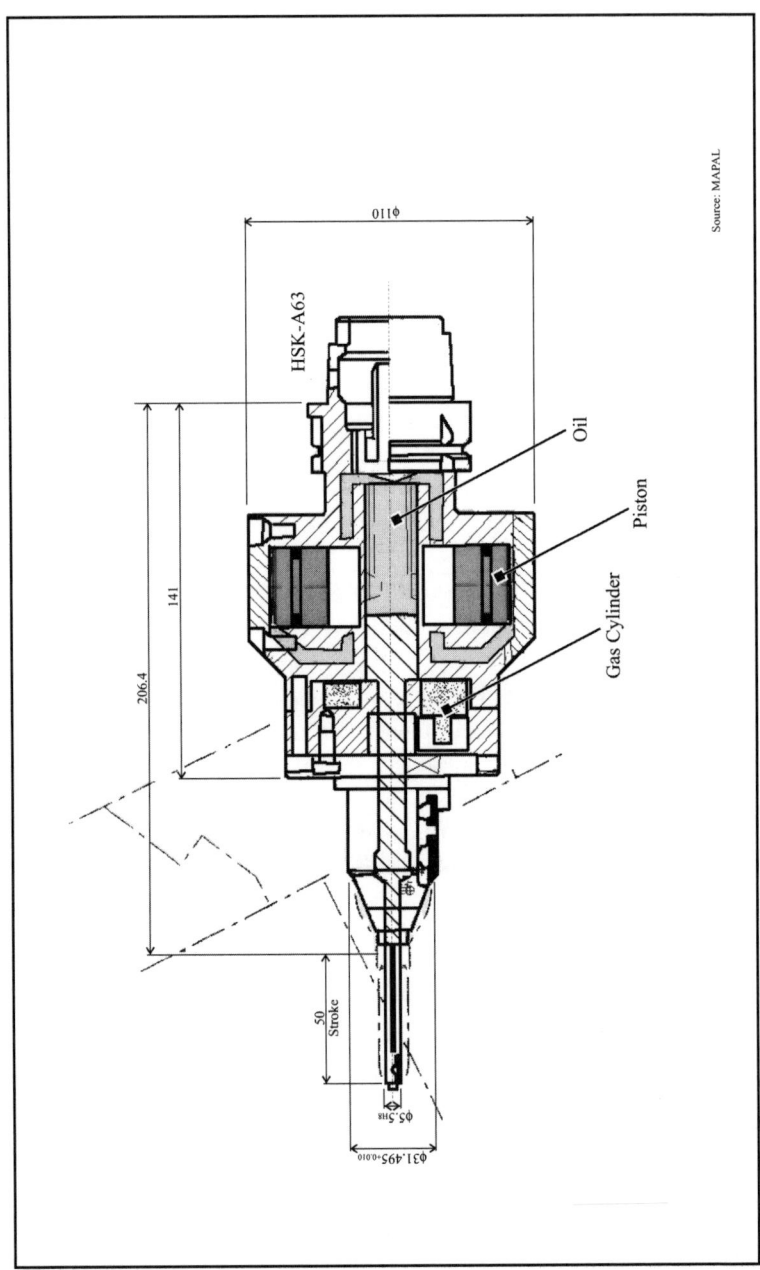

Figure 3.8 • High Precision Cutting Tool

tioning data. Thus, mechanical transmission errors can then be compensated for by the machine's control unit.

Machines can also be equipped with touch probes that perform automatic datum shifts, and eliminate possible positioning errors of the part to machine spindle centerline. There are also touch sensors that can locate the actual part position of a bore or a reference surface and trigger a corrective motion of the machine to ensure a precise machining pass. See Figure 3.9.

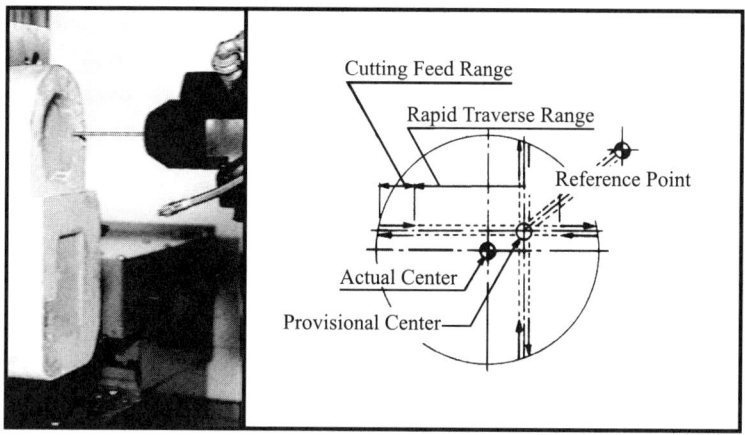

Figure 3.9 • Touch Sensor for Part Positioning

"Adaptive control" is a machine tool feature that continuously monitors tools during the cutting process. When the load on the machine spindle becomes excessive, and the sensor detects greater cutting resistance, the feed rate is reduced. When the load is returned to normal, it adjusts the original feed rate automatically or, depending on the programmed system, might even increase the feed rate if the load is too light. This is an important feature to assure consistent machining with predictable results, particularly when only one operation is used to finish the workpiece.

FINE-TUNING OF THE TOOL ASSSEMBLY

As machining becomes more complex, cutting tools must be more intricately designed, and "tool management" has to be more accurately defined. "Tool management" is well described in contemporary engineering literature. Suffice it to say that it consists of the tool's identification, storage, handling, transportation, availability, administration, and setting.

Let me extend the discussion of setting into "educated" and "dedicated" setting or "proactive fine-tuning." It starts with the absolute cleanliness of all individual components prior to their assembly. After assembly, the complete tool system has to be placed into a setting fixture or pre-setter. The setting device's function is to set the cutting edges dimensionally as predetermined by the blueprint. In high-precision manufacturing, this means setting within micrometers, and it should be done in a clean, dedicated area. To set tool systems with the highest possible precision, setting devices should feature electronic, digital readouts (easy to read), and possibly for settings in the vertical mode (weight). They should be equipped with ultrasensitive probes and/or an optical profile projector (accuracy). Setting fixtures play a pivotal role in the fine-tuning process, for they proactively (presetting) and reactively (feedback) set the desired tool dimensions.

The setting procedure itself requires setting the primary and secondary cutting edges relative to a reference point of the tool body. This includes step lengths and diameters. It is essential to proactively fine-tune them to one another, so that they stay within the close tolerance bands as per the drawing and/or process sheet. Often, the tolerances to be held by a one step tool compared to a multistep tool can vary greatly, and this has to be taken into account. Another important aspect of fine-tuning is to use only the tolerances needed.

Adjustable inserts can, within micrometers, be raised or lowered from a mean diameter to stay at the upper or lower limit of the tolerance band. Lowering the insert to the low limit can add to the stability of the system. Raising it can increase tool life by allowing for more insert wear. A complete cutting tool system should incorporate a component to correct its runout. A system, checked and adjusted in a setting fixture to true zero runout, will, invariably, show some degree of runout when mounted into the machine spindle. If the runout is excessive, the system can manually be "trammed" in through adjustable adapters to clock it as close to zero as possible.

Fine-tuning is usually done only once per system and spindle, due to the expected repeatability accuracy of the cutting tool system itself. Proactive fine-tuning, in principle, works only if the production floor provides feedback from machining to tool setting, and vice versa. Accurate setting does not permit "missing" of even 1 or 2 mm. Machining, in turn, has to make use of the optimum machining data. The inputs stimulate the proper design of the cutting tool system (cutting geometry, system rigidity, etc.) and the machining peripherals such as coolant passage, or chip disposal. Given the sensitivity of manufacturing within fractions of millimeters, proper handling and care of the cutting tool system as an assembly, and its individual components, must be assured within all areas of application.

Proactive fine-tuning essentially means to do it "right from the start," since any "work in process" taken away from manufacturing for corrective action is costly. Human error and machine defaults, however, cannot be ruled out. It is, therefore, absolutely necessary to have two or three sets of cutting tool systems ready for machining: one currently in use, one as backup next to the machine, and preferably one more in the setting area. Proactive fine-tuning also answers the important call for predictability of machining results and tool life.

Manufacturing needs to be confident that the workpiece will meet the geometries and finishes as expected at the end of the machining process. Accurate setting of tooling systems with high repeatability will help meet this requirement and will also allow for predetermined tool life. By preventing possible rework or scrap "at the end of the line," proactive fine-tuning is an important element of precision, one-pass machining.

SUBSTITUTING PROCESSES

Advancements in metallurgy and cutting tool technology have provided new opportunities to replace conventional operations with advanced processes. The objective is to shorten the time it takes to manufacture products, and to eliminate costlier operations. An equally important goal is to meet blueprint specifications with regular production machines. It is understood that the machining results are to be equal to or better than the results achieved with conventional processes. Figure 3.10 shows ranges of surface finishes that can be achieved with the respective machining processes. It also reveals some interesting overlaps, and demonstrates how attractive certain process substitutions can be.

Manufacturing decisions are made in order to select the machining

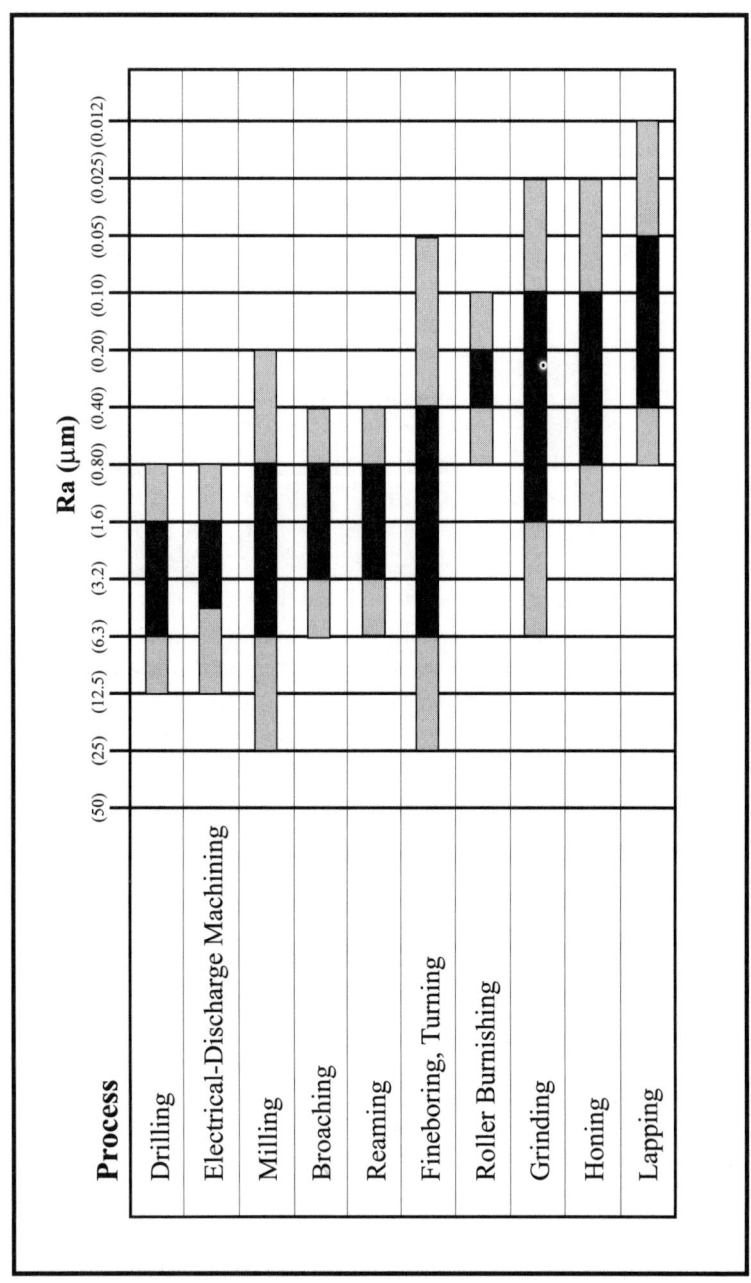

Figure 3.10 • Surface Finishes by Operation

operation which is most applicable for the overall production flow, and which can achieve the blueprint tolerances most productively and economically. Process substitution is invariably done with one-pass machining in mind. A classical case in point is substituting turning-hardening-grinding with hard-turning. Another recently developed process to replace turning and grinding is called O.D. reaming, illustrated in Figure 3.11. There are many other opportunities for substituting processes, including those shown in Figure 3.12.

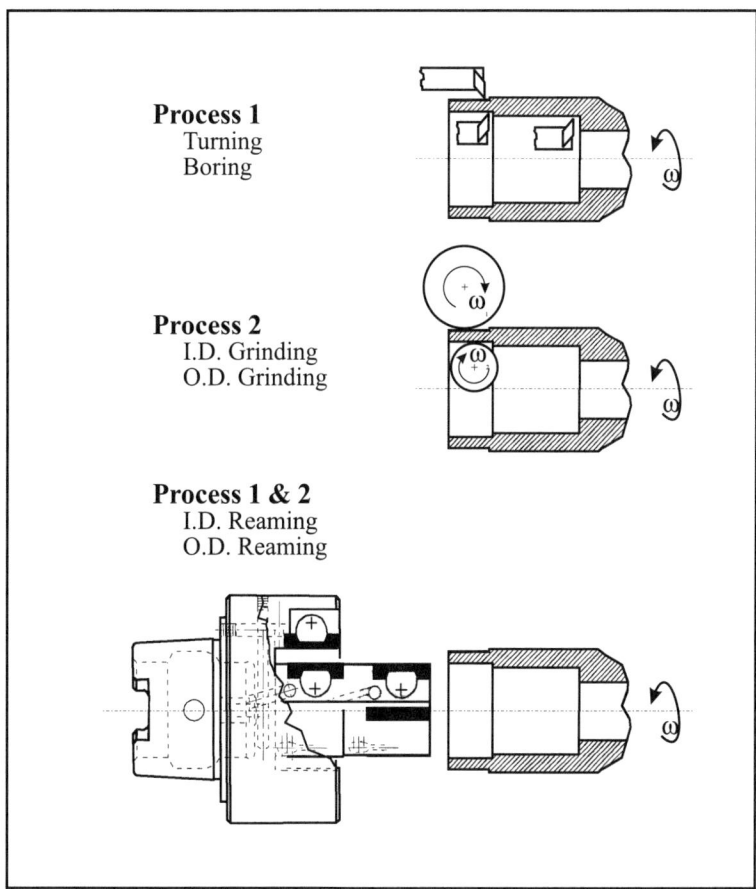

Figure 3.11 • O.D. Reaming

92 • Chapter 3

Replacement Process		Original Process	Reason
Milling	*Replaces*	Turning	One Machine For All (Other) Operations
Circular Milling	*Replaces*	Fineboring	Tool Flexibility
Broaching	*Replaces*	Hobbing (Shaping)	More Workpieces With One Pass
Electrical-Discharge Machining	*Replaces*	Slot Milling	Accuracy
Machining	*Replaces*	Reaming and Roller Burnishing	One-Pass
Fineboring	*Replaces*	Threading	Tool Flexibility
Thread Milling	*Replaces*	Lineboring	Accuracy, Machine Flexibility
Gunboring	*Replaces*	Dedicated Machine Station	Machine And Tool Flexibility
(Flexible) Combination Tools			

Figure 3.12 • Process Substitution

APPLICATIONS IN PRACTICE

Hard-Turning

Finish-machining of hardened ferrous material is still performed mostly by grinding operations, although the technology to machine those parts more economically and productively by hard-turning is available. While grinding usually requires several setups, hard-turning can be done in one setup and pass with dry machining. However, an absolute stable and rigid process has to be provided — small amounts of material removal as well as low feed rates reduce the cutting load. The cutting material is either polycrystalline cubic boron nitride (PCBN) or ceramics (Al_2O_3, Si_3N_4). With hard-turning, a large corner radius generates a limited amount of chips and creates a smooth surface finish. The cutting forces are high in conjunction with extreme high thermal stress. See Figures 3.13 and 3.14.

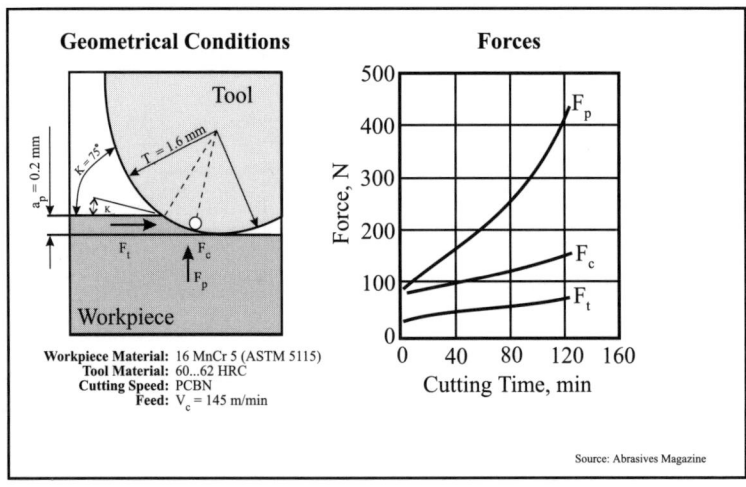

Figure 3.13 • Geometrical Conditions and Cutting Forces

Chapter 3 • 93

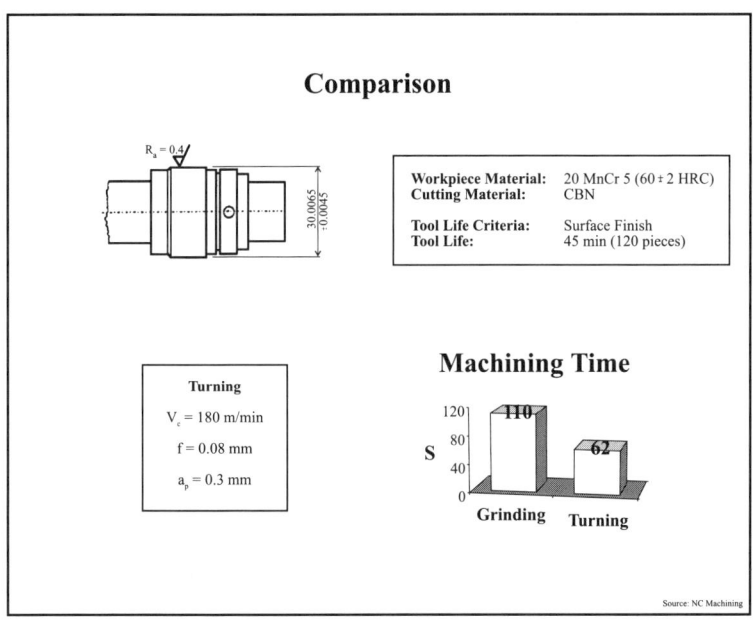

Figure 3.14 • Hard-Turning versus Grinding

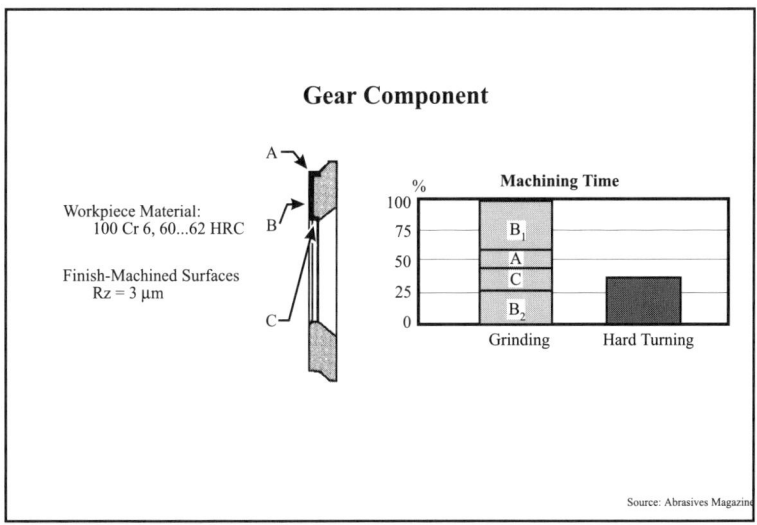

Figure 3.15 • Hard-Turning a Gear Blank

The three relevant criteria of one-pass finish-machining are machining time, surface finish, and cost. For example, finish hard-turning a gear component of approximately 62HRC hardness with cubic boron nitride cutting material demonstrates (as shown in Figure 3.15) the reduction in machining time in one setup, one pass, compared to a combined soft-turning/grinding operation, even when using a rather conservative cutting speed and feed rate — a tribute to reducing tool stress and thus tool cost. A comparison of the total production cost for finish-turning the gear component shows the cost for the tool itself to be almost 50% of the overall cost, as shown in Figure 3.16.

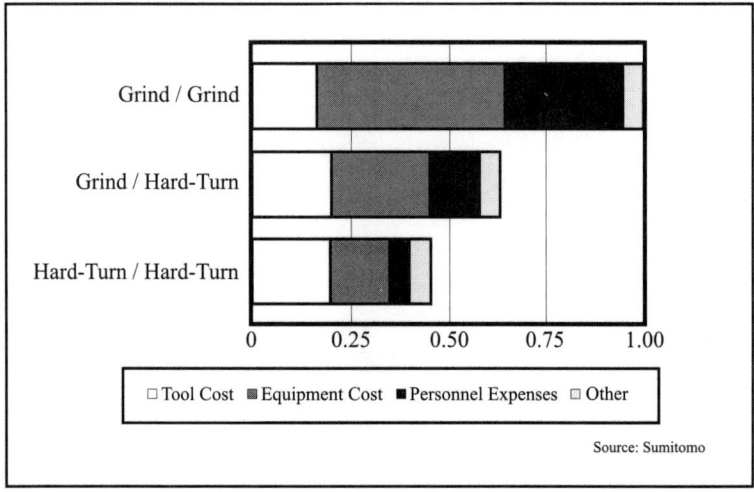

Figure 3.16 • Cost Comparison of Turning versus Grinding

Cutting inserts with multiple CBN corners make cutting material costs even more economical. For workpieces without interrupted cuts, and with lower batch production, the use of ceramics can lower the cost even further, as shown by:
• short machining times,
• flexible tooling to accommodate varying part design,
• machining on flexible, regular production machines,
• minimum production cost,
• no provision for coolants, and
• reasonable investment cost.

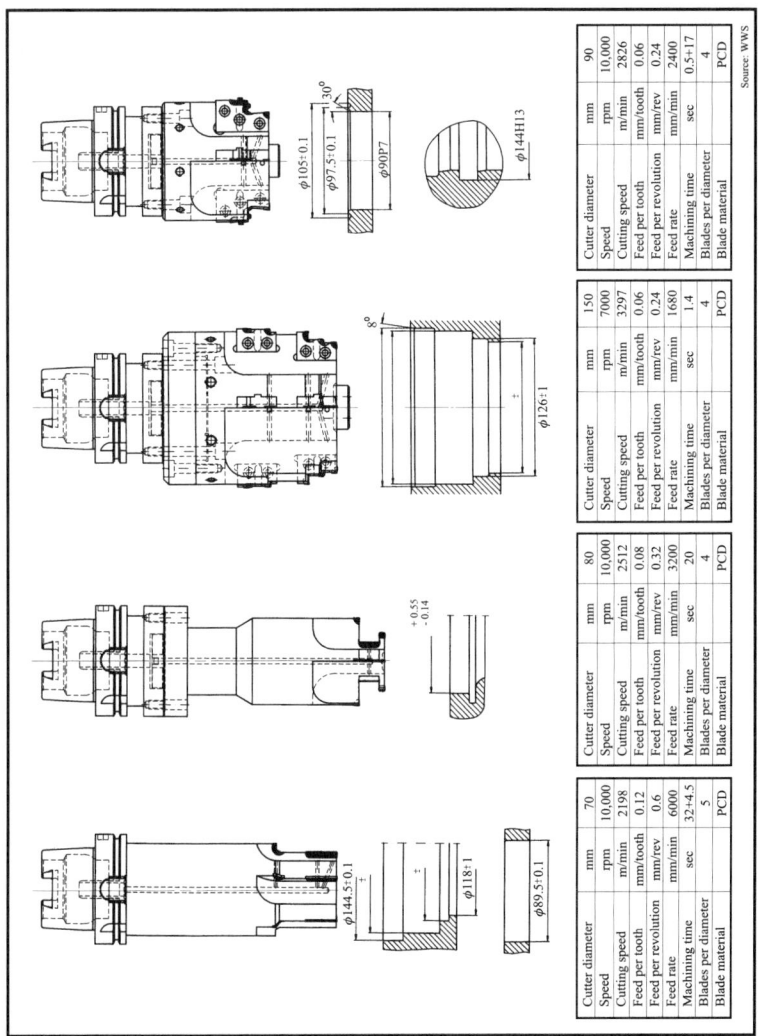

Figure 3.17 • Circular Milling Tools

This proves that one-pass, one-setup hard-turning is productive as well as economical.

Circular Milling/Circular Interpolation

Complete-machining of complex parts involving surface milling, boring, recessing, and retract-machining at high cutting speeds is per-

96 • Chapter 3

formed by circular milling tools. These tools machine larger aluminum workpieces such as gearboxes, transmission cases, crankcases, etc., at high cutting speeds and in less main machining time, while minimizing nonmachining time (setups, tool changes). They feature multiflute, fixed PCD-cutting edges ground to high precision, and a rigid, well-balanced design, including large diamond areas and precision shank configurations. See Figure 3.17.

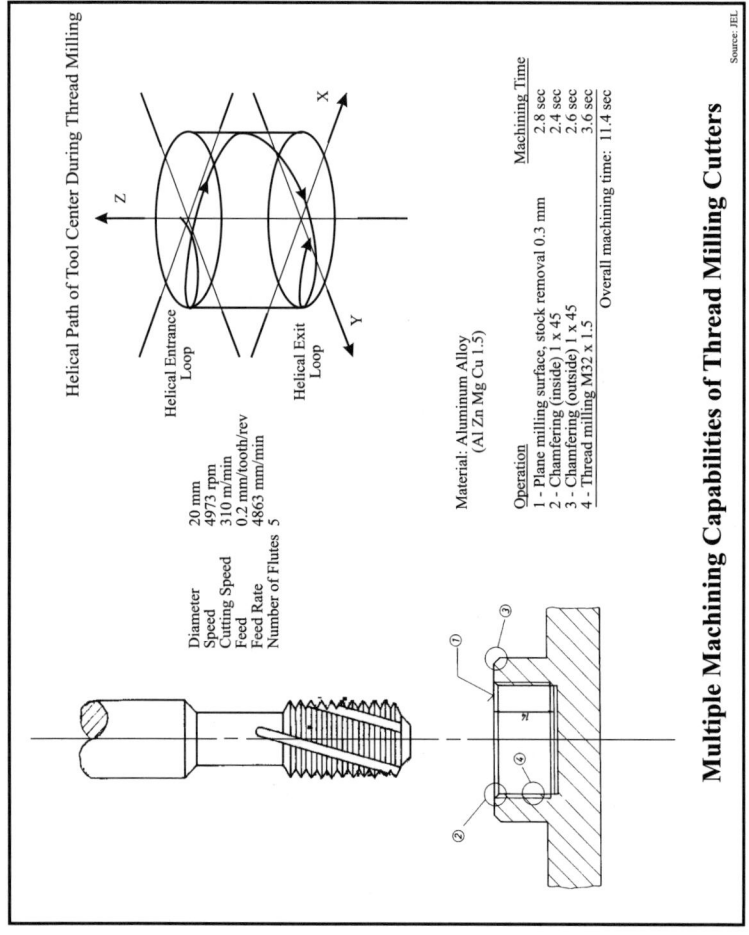

Figure 3.18 • Thread Milling

Circular milling tools can also be designed for interpolating bores. Perfectly round bores are usually made with precision boring or reaming tools which rotate around their own centerline. Interpolating tools, by contrast, incrementally generate the bore by describing a circular path until a round bore is achieved. This further minimizes tool quantities, for one tool can machine several bores of larger diameters, each in one pass.

Using the sophisticated control system of machining centers, programming machine tools to interpolate is a smart way to do separate operations in one setup with only one tool. Thread milling is a good example (see Figure 3.18). One tool, programmed for helical interpolation, chamfers, counterbores, and thread mills and can perform the operation over a wide range of diameters. Helical interpolation provides interrupted cuts, breaking chips easily and preventing "birds nests" when machining steel and aluminum. And, since the tool does not fill the hole, chip removal is improved. Furthermore, since cutting edge contact with the workpiece is as little as 30° per revolution, the temperature during cutting is greatly reduced. Circular milling and milling through interpolation have broadened the range of one-pass machining.

Fineboring

About 70% of all chip-making machining involves hole-making, and a high percentage is finish-machining closely toleranced bores.

Bore finishing is crucial because bores must receive rotating parts, bearings, seals, cylinders, valves, or other critically toleranced parts and components. Finishing multibore configurations consistently and predictably in one pass while maintaining IT5/IT6 tolerances with stringent surface finishes is no easy task. Given accurate, reliable, rigid machine tools, the key technology lies in meticulous tool design. An important criterion for multistep bores is the coaxiality of the bores and their defined dimensional interdependence.

These fineboring tools feature high-precision indexable cutting inserts and peripherally arranged guide pads to guide the tool through the bores and bridge possible air gaps (bore interruptions). See Figure 3.19. In aluminum, inserts and guide pads are made of PCD material to prevent material buildup and to extend tool life. Designed with a reliable shank system, e.g., HSK, and run on a quality-built machine, such fineboring tools reportedly machine precision bores within 5-digit part counts before changing the inserts.

98 • Chapter 3

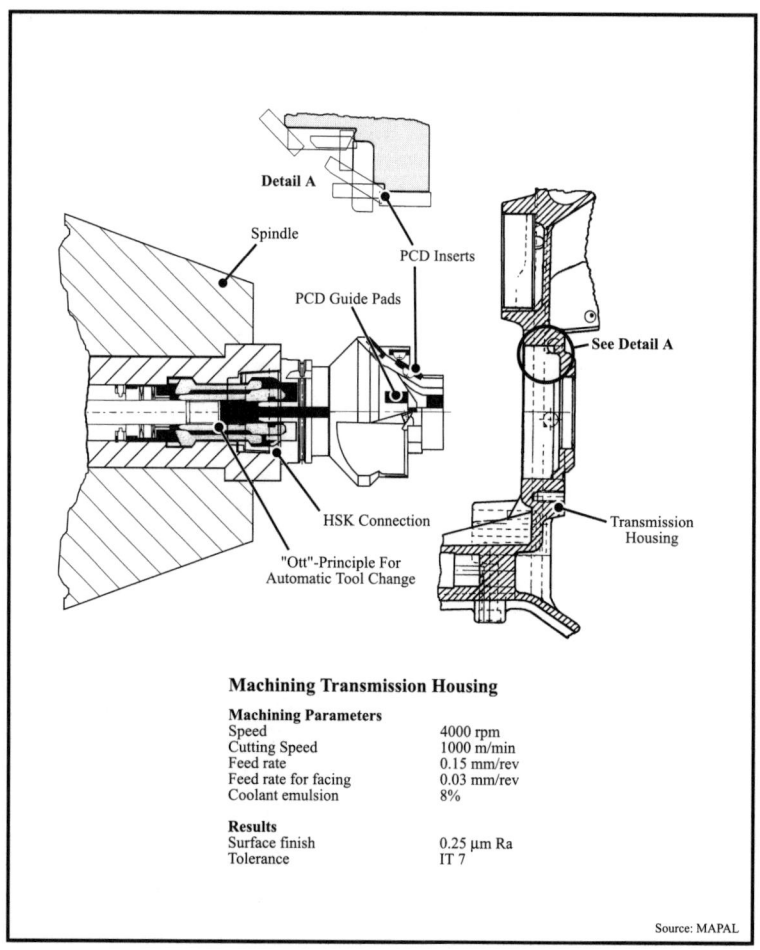

Figure 3.19 • Precision Fineboring Tool

In cases where the machine spindle is not as precise as necessary to produce the desired bore accuracy, in-between adapters that can be adjusted radially or angularly can tram the tool to "zero" runout. The described fineboring tools can run at very high speeds, do not require any bushing support, and are truly used as one-pass cutting tools that can finish-machine extremely demanding bore configurations, often without premachining from the cored bore.

Advanced Generating

When different part features are directly related, the complex workpiece contours have to be machined to incorporate the same related part interface within very close tolerances. Examples are the valve seats/valve guides used in engine cylinder heads (plunging and reaming), or the cylinder sleeve that is inserted into an engine block (boring and chamfering). Sometimes part surfaces have to be machined under circumstances where the tool's access is rather limited and the preceding area is dimensionally related. Examples are crankshaft bores (boring and facing) and differential cases (boring and milling). In these instances, so-called generating or feed-out tools have to be designed. Their use is not new. What is new, however, is the shift to flexible manufacturing, higher workpiece complexity, quality, and higher cutting speeds. To accommodate all demands into one cutting tool and one machining process, new design and engineering approaches have to be taken by using the laws of physics, hydraulics, and mechanics. Feed-out tools running on flexlines/cells can be actuated by centrifugal forces induced either by high speeds, the machine's coolant system, or well-balanced eccentric systems.

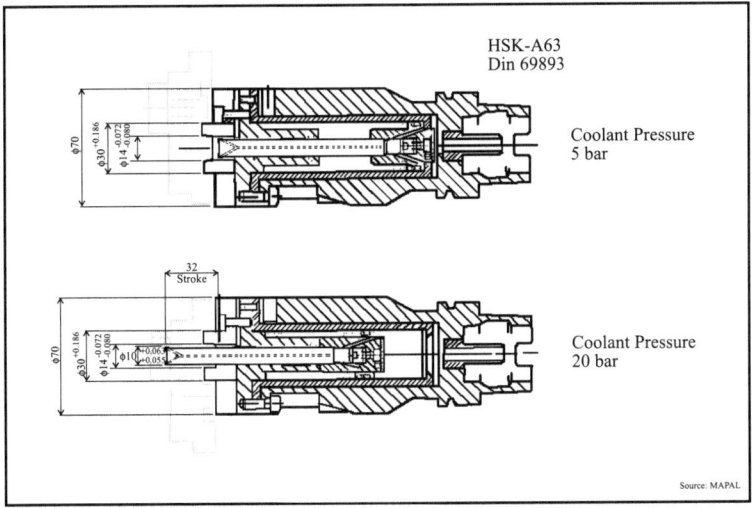

Figure 3.20 • Coolant Actuated Feed-Out Tool

The tool design illustrated in Figure 3.20 combines turning, boring, and plunging into one tool, and can run on a standard machining center. To finish-machine an aluminum pump housing, the tool will operate at two different coolant pressures: at 5 bar, it plunge-turns and faces; and at 20 bar, the smaller boring tool moves out of the large tool body to finish-bore a 10 mm bore. Featuring the accurate HSK-shank and precision-ground fixed PCD cutters, this process takes place at high cutting speeds and ensures excellent surface finishes and extremely long tool life.

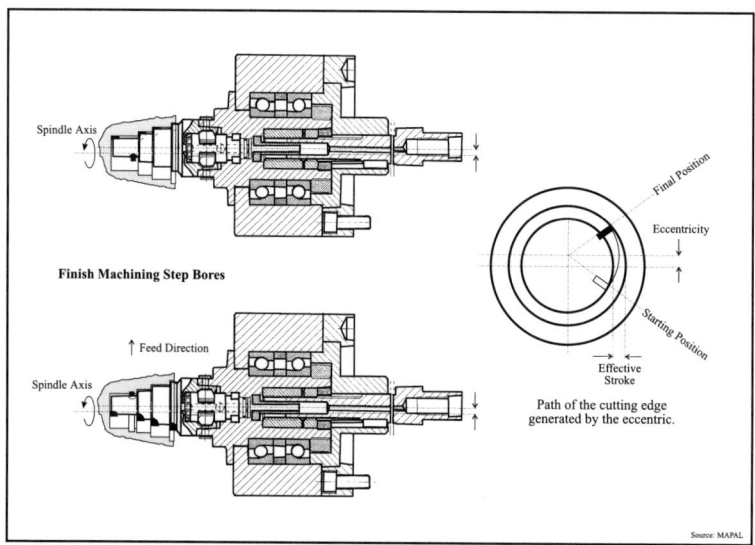

Figure 3.21 • Eccentric Actuated Feed-Out Tool

The tool shown in Figure 3.21 combines multistep bore finishing with facing and grooving operations at high cutting speeds. It features a boring bar with an HSK-interface mounted in an eccentric actuating head with a 160 mm diameter and a 10 mm stroke. Applied at 10,000 rpm, it runs at about 1,500 m/min cutting speed. The head functions by pulling the drawbar in the axial direction and by spiraling the round slide when rotated by 72°. Since the round slide is off center, crossfacing takes place.

To machine at IT7 tolerance consistently, quality balancing of the tool is crucial. First, the actual tool is balanced around its HSK-axis, then mounted in the eccentric head, and then the entire assembly is balanced. Possible out-of-balance corrections are done on the head.

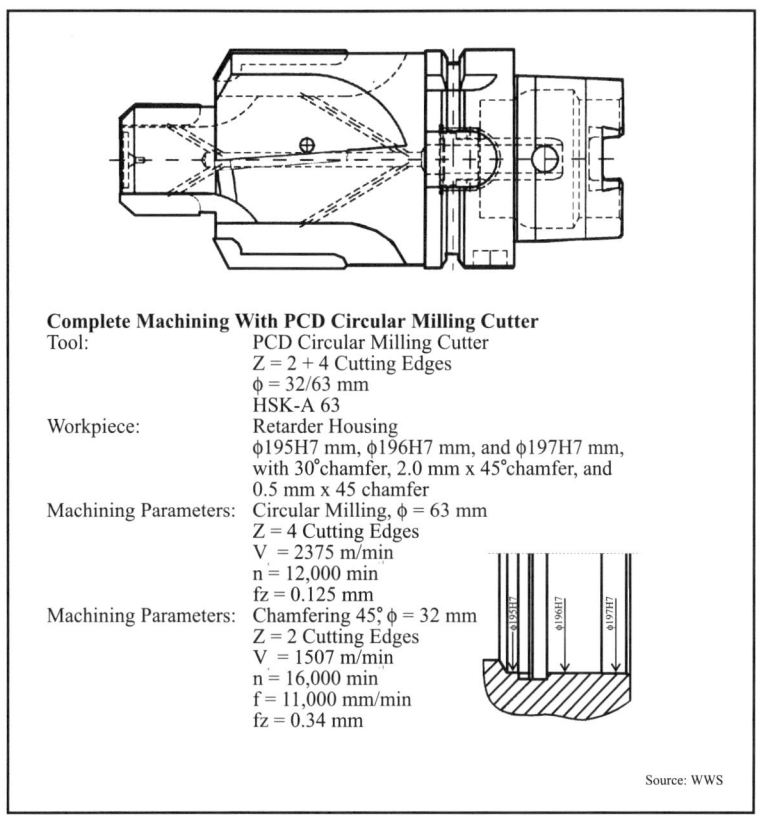

Figure 3.22 • Feed-Out Tool Actuated by Centrifugal Force

Another intriguing design, shown in Figure 3.22, is to use the centrifugal forces of high speed machining for feed-out motions and one-pass machining. At regular speed (about 1,000 rpm), the valve guide tool stays retracted in the tool body that plunges the valve seat. As the speed increases to 5,000 rpm, the centrifugal forces actuate the axial motion of the guide tool. This is done through two pistons, which push oil through a valve onto an axial piston providing a controlled feed rate for the tool. During that motion, the piston is compressing a spring that also moves the tool back into the retracting position as soon as the original (normal) speed of 1,000 rpm is reached.

Manufacturing needs to take this golden opportunity to eliminate unnecessary roughing and auxiliary postmachining operations and combine step-by-step machining procedures into one machining pass. The demand for weight reduction and higher quality in most industrial manufacture calls for more accurate raw castings with no excess material and no impurities. Investment casting is becoming more popular because it allows the widest net part size range of any casting process for thin walls, undercuts, and internal passages.

In forging, too, near net-shaped parts are formed by grain flow optimization. It increases impact strength and fatigue endurance limits, thus allowing for thinner wall design and the reduction of excess weight and stock, thereby allowing for the diverging criteria of light weight and high strength. Proper design and material selection can accommodate both criteria as, for example, in the automotive industry where aluminum is used for engines and transmissions.

As design and manufacturing engineers try to squeeze more functions into smaller parts and subassemblies of sophisticated end products, the parts become more complex and difficult to machine. These lightweight, accurately cast, complex workpieces of advanced materials invite one-pass (precision) machining for even the most demanding applications.

One-pass machining needs to be pursued by manufacturing because it

- eliminates stacking up tolerances (step by step machining),
- decreases nonmachining time (setup, tool changes),
- decreases main machining time,
- lowers tool inventory,
- promotes lean manufacturing,
- accelerates the manufacturer's response time to market shifts and changes, and
- complements high-speed machining and facilitates (near) dry-machining.

Chapter 4

GLOBAL CONCURRENT MANUFACTURING

TIME, QUALITY, AND COST

Ever since Adam Smith divided labor in his "18th-century factory" in order to allow each worker to do just one step of the assembly process, speed has been an issue in manufacturing. In the 1990's — and well into the next century — speed is simply a business issue. Manufacturing companies have been forced to speed up their processes, operations, development cycles, and their responses to customer orders. In short, the modus operandi is time-to-market, rapid implementation of automation, and quick response to changes, which requires speed, agility, and synchronization.

Besides speed, it is quality by which corporations are measured. The adoption of quality procedures defined by Deming and Taguchi has catapulted the quality of products to ever higher levels, and there is no end in sight. What counts in the marketplace is not only the quality of the product itself, but also the reliability, credibility, and competence of the supplier. These are equally important performance measures in the eyes of customers, thus triggering the never-ending quest for continuous improvement.

While the buyer's intention is to continually get more features out of the product purchased, price is an equally important factor in the decision to buy. Providing more, possibly at less cost, has led manufacturing companies to search for intelligent decisions on the process level in terms of product cost, make or buy, just-in-time delivery, and plant locations. In addition, the principles of Cost to Design (CTD) and Design for Manufacturing (DFM) are part of the cost equation on the process level. The performance measures of time, quality and cost are at the core of any successful manufacturing company. See Figure 4.1.

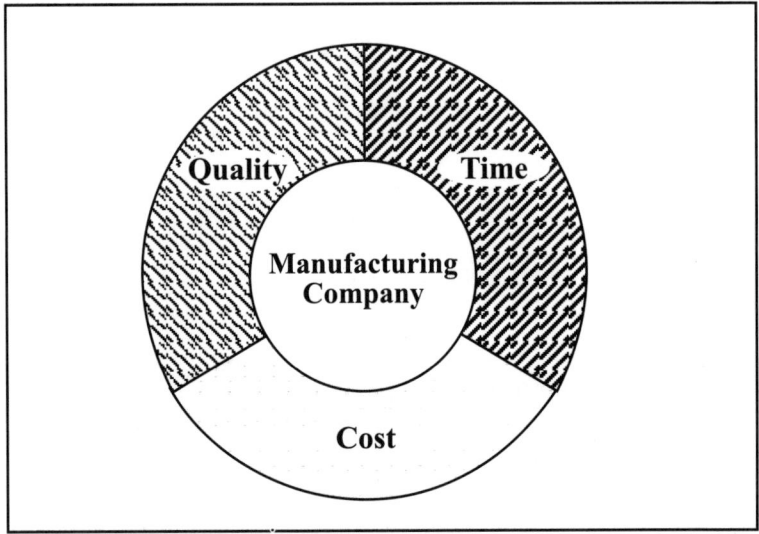

Figure 4.1 • Benchmarks - Time, Quality, and Cost

THE RELEVANCE OF CONCURRENT ENGINEERING — ITS PURPOSE AND PRINCIPLES

Today's manufactured products are extremely complex due to health and safety features, a whole array of possible options, and computerized technology. Any company that operates on a worldwide basis expands on complexity per se, because of challenges in procurement, product uniqueness, pressing lead times, and the relationship with suppliers and distributors (coalition), partners, and multifaceted clientele. Management systems must be powerful and coherent to meet those challenges by using vehicles for pooling collective corporate wisdom to yield the best results at the least cost.

The principle of Concurrent Engineering (CE) has a proven record of dealing with corporate challenges. By stressing a team-oriented collaborative approach to product design and processes for manufacturing, CE is ideally suited for product development and product improvement because it has a positive impact on time, quality, and cost. It minimizes:
- lead times,
- design changes at the production stage,
- the use of critical processes,
- unit costs,
- variability, and
- unrealistic tolerancing.

It maximizes:
- standardization,
- supplier knowledge,
- process predictability,
- computerized communication,
- the use of advanced technology, and
- design simplicity.

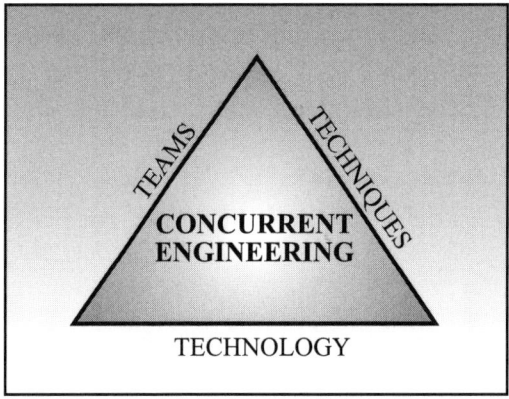

Figure 4.2 • Concurrent Engineering Trilogy

Concurrent engineering can best be described as the synthesis of teams, techniques, and technology. Technology includes the knowledge and availability of the systems and processes needed to produce the product (see Figure 4.2). There are many techniques or subsets used in the process of CE, including:
- engineer to manufacturability (process capabilities),

106 • Chapter 4

- engineer to reliability (e.g., maintenance, product life cycle),
- engineer to cost (e.g., product cost, market price),
- engineer to application (e.g., consumer product or production product), and
- engineer to quality (SPC, TQM, etc.).

The team concept is at the heart of CE. Its power is the pooling of professional backgrounds, skills, knowledge, expertise, and the application of problem-solving techniques. The team is committed to a common purpose, guided by a set of performance goals and by an approach for which they hold one another accountable. In CE, the team shares leadership roles, encourages open discussions and sensible decisions, and measures performances collectively. Competence and commitment strengthen over time. Teams go through stages (see Figure 4.3) starting with information gathering, typically marked by uncertainty, followed by the "storming stage," where the individual team members jockey for position and compete for power or attention. The team then enters the "norming stage" where they establish cohesiveness, share responsibility, and gain confidence. Finally, they reach the "performing stage," with a strong sense of team identity, cooperation, and effectiveness in goal achievement. It is important that adequate time is allowed for the team to mature.

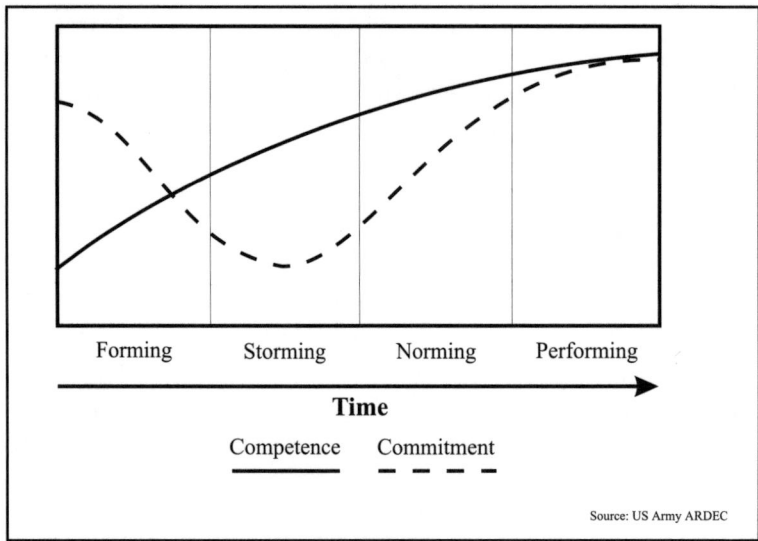

Figure 4.3 • Stages of Team Development

Who should be on the team? Essentially, the team should include representatives of relevant departments including, but not limited to, design, engineering, manufacturing, product support, purchasing, sales, quality assurance, and production planning and processing, as well as critical suppliers. Such multifunctional, multitalented teams have the power to inject divergent approaches into the discussions. By taking into account the company's interdisciplinary needs and functions, the team can consider such issues as the launch of a new product line, the change of a production process, the development and implementation of more versatile production technology, the measurement of a current product's competitiveness or process performance, and how manufacturing productivity will be impacted by anticipated design changes. It can also initiate research and development for new products, manufacturing processes, or organizational restructuring.

Product Life Cycle

When should the process of concurrent engineering begin, and when should it end? A manufacturing company seeking an opportunity to introduce a (new) product, must begin with team-building and must take steps to assure that the team becomes a permanent fixture within the organization for the product's whole life cycle — from concept to its market exit. The concurrent engineering process is then renewed for yet another product or variation of the existing one, allowing the corporation to follow a defined road map for the product's life cycle and the respective concurrent engineering process. See Figure 4.4.

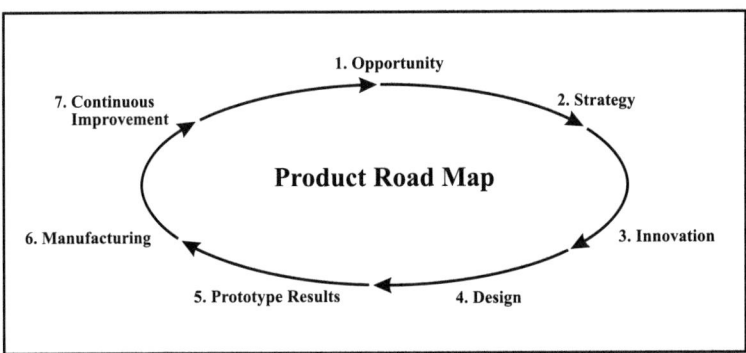

Figure 4.4 • Product Road Map

108 • Chapter 4

Opportunity

Any manufacturer who is in the market for the long haul has to be fluid and flexible, i.e., able to adjust to market gyrations and the erosion of its products. It must put itself into position to uncover and take advantage of hidden opportunities. It must know its capabilities and abilities, and realistically identify the competition within its business environment. The team's first meetings will be purely fact finding missions to identify potential opportunities.

Strategy

Against the backdrop of what is good for the company and the customers, a precise strategy has to be defined. The following questions must be dealt with. How do we position ourselves with the product? What do we offer with our product that others don't? How do we produce it, and what design principles are to be applied? Generally speaking, all questions ask "what directions do we take?"

Innovation

Benchmarking and the use of proven practices are the tools of the innovative segment of the product road map. The common denominator of the group's activities is imagination, and anything is possible at this stage. All combinations of material, tool, process, and technology ought to be evaluated to determine the best possible integration of processes.

Design

The product road map shown in Figure 4.4 illustrates the importance of the team's involvement in *product design* and in *design techniques*, for it is here that the success or failure of the product is principally determined. Design costs typically run approximately 5% of a product's cost, but design can account for 70% of its success (see Figure 4.5). Innovative ideas have to be implemented skillfully at the design stage. The team's awareness of solving problems early in the design stage cannot be overemphasized. The product's cost, quality, and functionality, as well as its manufacturability, are determined at this stage.

Prototyping

The manufacture of the first prototypes (forerunners of the final product) will reveal design strengths and weaknesses. This is the time to measure the results of the team's efforts up to that point. It might become necessary to draw up design improvement plans. Any adjustments

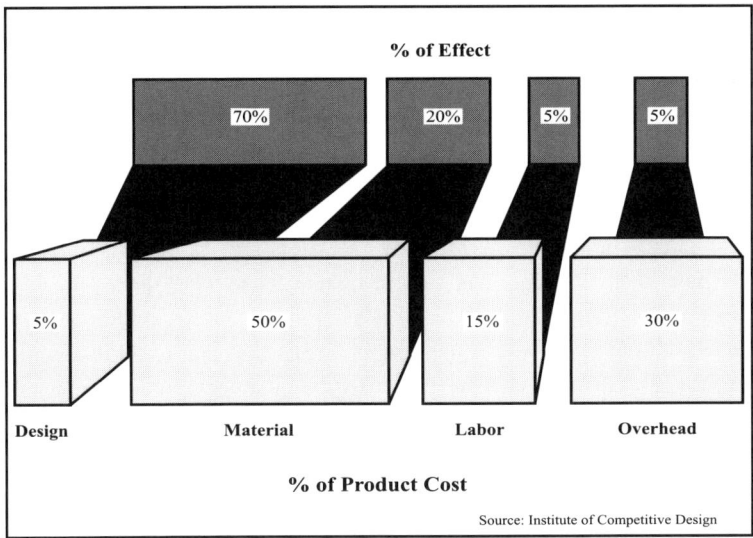

Figure 4.5 • Percentage of Product Cost and Effect

to the product or the manufacturing process must be undertaken at this point before full-fledged production commences.

Manufacturing

Once the go-ahead for manufacturing the product in its existing form is given, product changes will become impossible, impractical, or very costly. In fact, today's high quality manufacturing pursues the principle of "first part, good part," leaving absolutely no room for any adjustments. The concurrent engineering team has to take responsibility for smooth, productive, and efficient manufacturing.

Continuous Improvement

Product development is the core task of concurrent engineering, for it turns concept into product. However, this is not the end of concurrent engineering. *wt-Production and Management* magazine (Springer-Verlag) states that 75% of all errors are made during product development, but 80% of all errors show up at the end of the value-added process. The principle of continuous improvement must be firmly embedded in concurrent engineering. Core teams must deal with whatever improvements can be made. In this lies the continuum of concurrent engineering — an ongoing process for managing technology.

Applied Concurrent Engineering: Leading The Way To Concurrent Manufacturing — A Case Study

Saturn — A New Approach to Manufacturing

Eight years after launching the concept of a new division, General Motors unveiled its new Saturn car in October 1990. Manufactured in Spring Hill, Tennessee, Saturn subcompact cars are GM's answer to "the Japanese imports." Saturn's stated mission is to "market vehicles developed and manufactured in the United States that are world leaders in quality, cost, and customer satisfaction through the integration of people, technology, and business systems, and to transfer knowledge, technology, and experience throughout General Motors."

Meeting the self-imposed challenge of being the best in the industry calls for acquiring and applying state-of-the-art technology, processes, systems, and methods. After having spent approximately 4 billion dollars in developing costs, facilities and machinery, Saturn has become a full-fledged, self-contained car manufacturer. Its vehicles are available in two versions of a four-door sedan, a station wagon, and a two-door sports coupe with a 1.9 liter engine either with 64 kW engine output or an upscale version with two overhead cams and 4 valves per cylinder, delivering 92 kW output. All vehicles are available with either a five-speed manual or a four-speed automatic transmission. The demand for the Saturn car is high, and demand has exceeded supply. Saturn's remarkable success is, to a great extent, due to different manufacturing techniques, most notably the production flow of subassemblies, the balance of flexible and dedicated tooling, the lost foam foundry, and the engine and transmission lines.

Production Flow of Subassemblies. There are three operations — body systems, powertrain, and vehicle systems. In body systems, the car bodies are stamped, fabricated, and painted. The products flow south. The powertrain incorporates the foundry, casting, machining, and assembly of completed engines and transmissions. These products flow north. Both meet in the middle in vehicle systems, where assembly of the end product takes place.

Balance of Flexible and Dedicated Tooling. Machining of the powertrain components is done by flexible or dedicated machines, with standard and advanced cutting tools. Flexibility for product mix and changes in parts requires special and dedicated machines for both dedicated and fixed part runs.

Lost Foam Foundry. The lost foam casting operation supplies the engine block, the cylinder head, the crankshaft, and the differential case. Saturn is the only car manufacturer to use this casting method in high-volume production. Its main advantages are design flexibility (component integration into the base casting), near net-shape castings, and higher accuracy, resulting in fewer parts and machining passes.

Engine and Transmission Lines. Automatic and manual transmissions are built on the same assembly line. Up to 75% of either type can be produced at the same time, depending on demand. The engine assembly line offers the same flexibility for building either the standard or high output engines. Saturn, to date, is the only U.S. car manufacturer with such optional flexibility.

Saturn's Applied Concurrent Engineering

I became part of GM's Saturn project as an "outsider" brought in to inject expertise in finish-machining and advanced cutting tool systems for manufacturing . This provided me with the opportunity to observe Saturn from the very beginning, when they consisted of a small task force, to today's truly remarkable independent car manufacturer. What Saturn stands for today is the product of concurrent engineering in its truest sense and clearest form. Saturn started with a clear goal.

Setting the Highest Standards. Meeting the self-imposed challenge to being the best in the industry, the task force approached key players in manufacturing to find out their technical and technological limits. They really went for the extremes in applicable machining data, material, automation, processes, and product support — the complete manufacturing envelope. They wanted to know where the ceilings were and whether, if necessary, they could go on from there. They gathered and collected all data imaginable from the respective industry leaders. After establishing standards and gathering data, they proceeded to the first step toward their goal.

Cross-Functional Teams. It was surprising to meet people at Saturn that I had previously worked with in other divisions of General Motors such as Pontiac, Saginaw Steering, GM-Tech Center, Cadillac, Oldsmobile, and others. Engineers from design, manufacturing, planning, engineering, service, R&D, processing, quality control, fabricating, metalworking, and assembly came to Saturn, where they were combined into cross-functional teams. There were even design engineers to accompany manufacturing engineers on business trips, and vice versa. Those teams developed into expert teams after they analyzed all the data they had previously gathered to conceptualize areas of manufacturing. Simultaneously, they benchmarked subassemblies, i.e., they acquired, for example, engines and transmissions of the leading manufacturers they considered to be the best in that market segment. They spread out all individual parts and analyzed them for form, fit, and function. Then they started preliminary designs, applying the best answers available through benchmarking, their own expertise, and outside (supplier) input. At that point, Saturn had reached the next stage.

Product Development. They began making models and prototypes, testing them, and verifying their performance. Soon they began selecting materials and processes. They drew up the plans for production and tooling, training, and product support. Decisions of whether to buy or build were made, and at that stage qualifying suppliers entered the picture. The overriding principles of all discussions of the product development phase were to:
- build a product that satisfies the market need for function and performance,
- satisfy the customer's perception of what is bought,
- give value for the money spent,
- secure long service life and ease of maintenance, and manufacture the product economically,
- use the best processes and tools,
- maintain consistent quality, and
- achieve the lowest cost.

The next step was to work with outside sources.

Supplier Integration. The involvement of key suppliers took place at an unusually early stage of the concurrent engineering process. It appears that the plan was to build a first and final product with key suppliers at hand, and then integrate general suppliers (a third supplier layer was installed shortly before completion of the manufacturing facilities). In order to accommodate the JIT-delivery principle and the "window"-supplier approach, Saturn established one primary cutting tool supplier, who sourced all tooling from other vendors. By doing that, Saturn established one outside counterpart to purchase tooling, leaving all direct contact lines to "sub"-suppliers open for technical questions and problems. Figure 4.6 illustrates the supplier integration process as it appears to exist at Saturn.

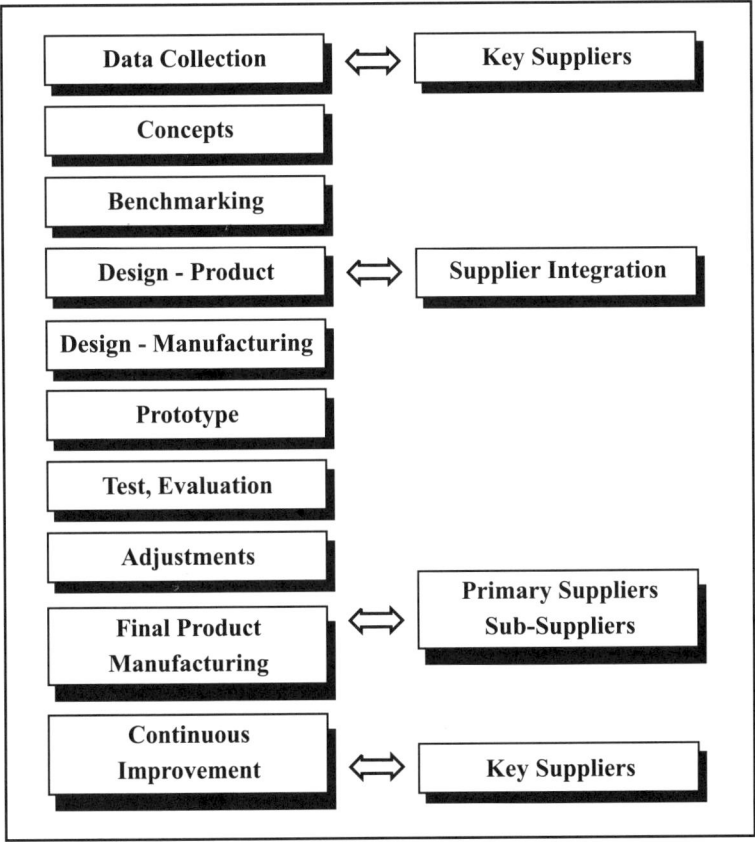

Figure 4.6 • Concurrent Engineering Process

Continuous Improvement. Ever since the inception of high volume part production for subsequent final car assembly, the process of continuous improvement has been integral to manufacturing. At Saturn it is done through "expert teams" with close ties to suppliers. That means concurrent engineering is an ongoing process. With teams in design and manufacturing working side by side, concurrent engineering is used at Saturn to detect and eliminate potential manufacturing problems early on.

Concurrent Manufacturing

Product development and product improvement, directed and orchestrated through cross-functional, interdisciplinary teams using the principles of concurrent engineering, have become invaluable tools for manufacturing companies. What is lacking, however, is the necessary continuation of this process on the production floor itself, particularly when manufacturing identical or similar parts or part families is done by several plants. Duplication of efforts, the artificial creation of "in-house" competition, variation in product quality, product cost, and extended time to market are costly results. These potential conflicts and impediments contradict the basic ideas of concurrent engineering and, in fact, can erase any achievements realized by successful teams. Ultimately, a manufacturing company can disqualify itself from becoming world-class and competitive on a global scale. As illustrated in Figure 4.7, applied concurrent manufacturing evolves from concurrent engineering and involves

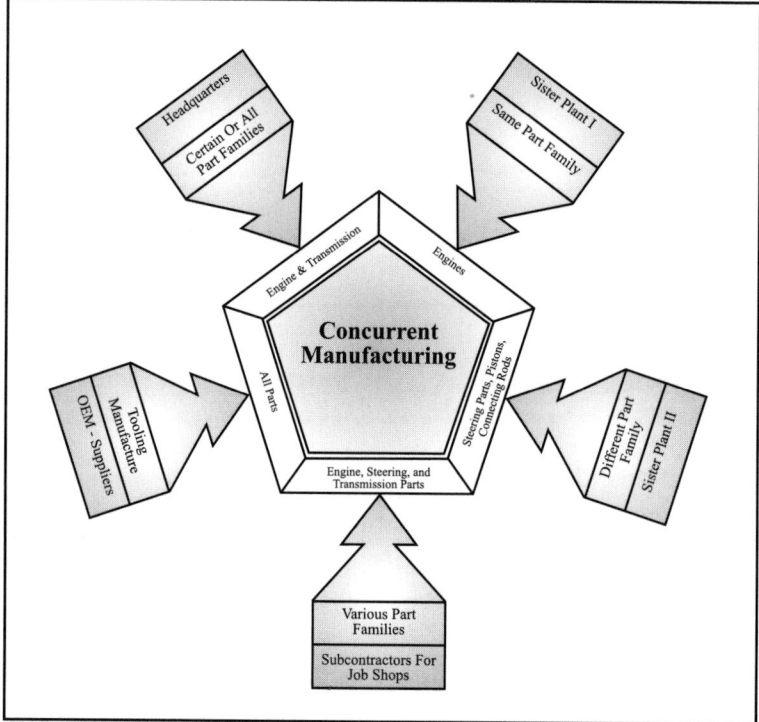

Figure 4.7 • Contributors to Concurrent Manufacturing

headquarters, sister plants, subcontractors (job shops), and OEM suppliers.

The Concurrent Manufacturing Team

To accommodate ever-changing, ever more intertwined and (sub)divided manufacturing responsibilities, interdisciplinary manufacturing teams have to be formed. This encompasses the gathering of ideas, feedback, and information, as well as identifying improvements to problems, events, and occurrences at the manufacturing level, with a link to concurrent engineering teams. The concurrent manufacturing team consists of process engineering, product engineering (the link to concurrent engineering), plant management (upper and lower), tool engineering, tooling suppliers, quality and inspection, research and development, and communication support, as shown in Figure 4.8.

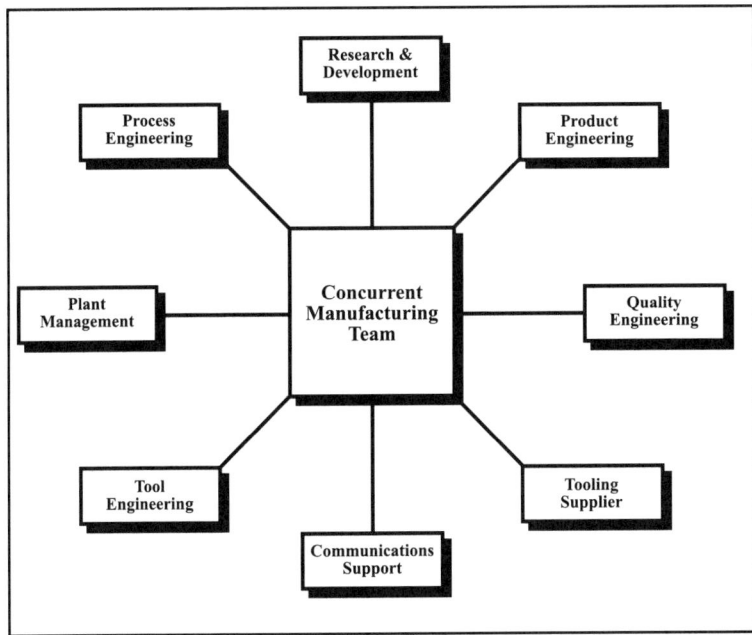

Figure 4.8 • Concurrent Manufacturing Team

Core Principles of Concurrent Manufacturing

The concurrent manufacturing team's focus primarily comprises the following:

Standardization. Uniformity of processes can more easily be achieved through pursuing high standards, both internal and external, for machining and manufacturing. External standards are issued by standards agencies and applied industry wide, e.g., ASME and ISO standardization of tool features. Internal standards are set by corporate manufacturing and apply to all contributing manufacturing plants (sister companies, subcontractors, etc.) and often include methods of process monitoring, inspection, machining, etc. Standardization is an important contributor to reducing manufacturing time and cost while increasing manufacturing transparency and easing the burden of communication.

Optimizing Processes. For concurrent manufacturing teams, the key is *process*. It is the method by which, and the system through which, products are manufactured. As Deming, Juran, and Taguchi stress, the focus has to be on the process, its stability and improvement. When manufacturing companies focus on *how* the product is made, they can continue to find ways of improving quality, cost, and time even after all product defects have been eliminated. Continuous improvement and process innovation really go hand in hand. (See Figure 4.9.)

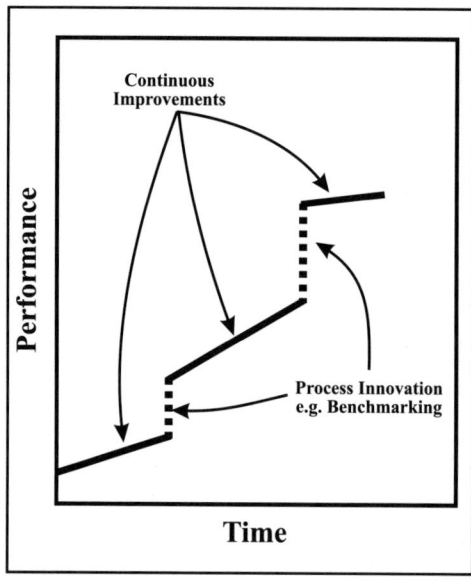

Figure 4.9 • Process Innovation for Continuous Improvement

Innovative processes can come about through R&D efforts, the input of a tooling supplier, or any of the other participating branches of the concurrent manufacturing team. They can then be applied to impressively accelerate performance on the production floor. The value of process improvement becomes obvious when describing the relationship of precision and performance; increasing the efficiency of both will result in the simultaneous increase in process improvement as shown in Figure 4.10.

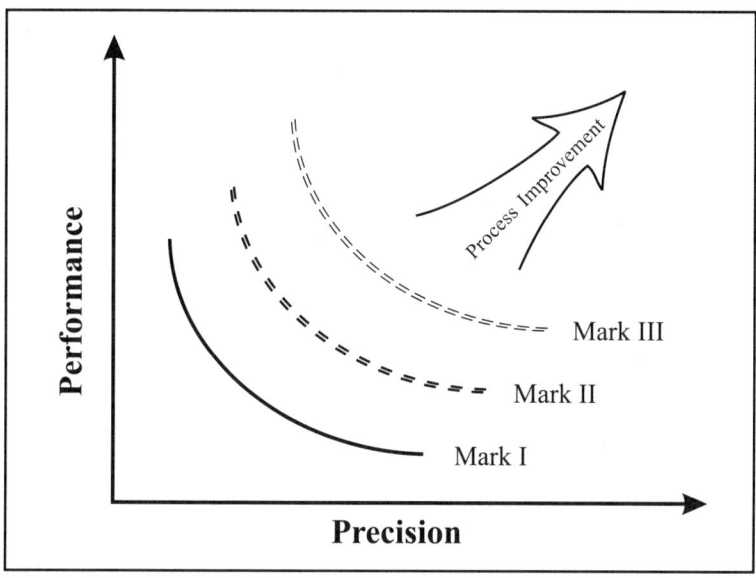

Figure 4.10 • The Relationship between Performance and Precision

Tool Systems. Machine tools and cutting tools usually represent the biggest investment on the production floor, and they often have the biggest effect on the process. Furthermore, they tend to affect most of the other elements and factors of manufacturing such as parts supply, transportation, monitoring, gauging, assembly, product throughput, etc. As a result, tool systems play an important role within the principles of concurrent manufacturing. Tooling improvements contribute markedly to a manufacturing company's success and competitiveness.

Improving Performance Benchmarks. Cost, quality, and time are the performance benchmarks. Focusing on performance and operation issues, manufacturing's goal is to be competitive and world-class. Each company must determine how its processes compare to others, and whether it is producing the desired products. To improve performance benchmarks, either internally or externally, it is always necessary to invite comparisons. Internally, when a company compares its processes within its own organization, it does so from plant to plant. Externally, it wants to find out how it fares against direct competitors or other industries.

Benchmark comparison times involve: production time, transportation time, preparation time, idle time, and auxiliary process time.

Benchmark quality evaluations use quality standards such as ISO 9000 to meet or exceed customer expectations. Other criteria, more specific to individual industries and in-house makeup, often supplement these standards.

Benchmark cost reviews are used to control the operational floor expenses through performance criteria such as production throughput, machining times, inventory levels and turnover, and production planning.

Trends driving the adoption of concurrent manufacturing are:
- outsourcing,
- uniform quality,
- offshore manufacturing,
- technical complexity, and
- data management.

Outsourcing. Clearly, one of the most effective management strategies is to purchase products manufactured by a third party to complement or complete one's own product. Called outsourcing, this tactic lowers the content of self-produced parts of a product made of an array of components. An example is the automotive industry, which is usually a harbinger of things to come. The consensus of the "Big Three" in Detroit is to have a healthy mix of parts made at their own plants and other parts made by subsuppliers.

Studies show that car manufacturers should produce about 40% to 45% themselves, and have the majority made by *bona fide* suppliers. The complexity of parts and subassemblies, changing markets, the rapid advance of technology, and the need for steady growth and lasting profitability are difficult to handle without outsourcing. Outsourcing alleviates the increased need for change and the need to acquire every new technology. It allows the company to concentrate on its core abilities and capabilities, and to adjust for seasonal ups and downs based on respective production needs. Outsourcing is a definite indicator of the corporate cultural change of how products used to be made, and poses new challenges to be managed altogether differently.

Certainly, products that are outsourced have to be of the same or better quality, they have to be produced with predictable quality at the time they are needed, and they must be available at reasonable prices. Controlling and orchestrating these new supply chains is no easy task. Forming so-called strategic alliances between the original manufacturer and its supplier is important, but it is difficult to have all manufacturing entities on the same level, on a day-to-day basis. The key is to use communication and networking strategies.

Concurrent manufacturing with built-in concurrent engineering is the strategy that allows for disciplined manufacturing by removing guesswork. Essentially, manufacturing processes have to be streamlined, meaning that the same technology can be applied at all levels of the organization, and any possible improvements can be implemented simultaneously on the production floor at the original manufacturer as well as at the supplier. In cases where the original manufacturer does not produce the same or identical parts, it is used to share the knowledge of processes and advanced technology. The contractual agreements should spell out the interdependence of teams directly on the production floor, enhancing the best possible manufacturing processes.

Quality Standards. Today, companies do not just make products — instead, companies build and sell quality. Meeting customers' requirements is the goal; exceeding them is a never-ending desire, hence the perpetual quest for continuous improvement. The truth is that the bars for high-quality standards are continually raised higher and higher. Mandatory SPC/CPK-based manufacturing demands reduction of variability. The inception of even the most complex new manufacturing system begins with the principle of "first part, good part." Once in full swing, the objective is to assure zero-defect machining and manufacturing.

Every manufacturing plant deals with their adopted quality procedures individually. TQM (Total Quality Management) places importance on preproduction processes such as prevention, built-in quality, and quality in every activity of an organization. QFD (Quality Function Deployment) carries the voice of the customers through the organization, stressing teamwork internally and externally, vertically and horizontally.

The stringent requirements for producing products has plants scrambling for help and assistance. Manufacturing quality circles of plants producing parts and subassemblies for one and the same end product (automobiles, aircraft, construction equipment, etc.) have to set the same quality standards, coordinate them with one another, and take <u>collective</u> corrective actions. Continually improving the quality of manufactured products has to be achieved by examining and improving performance levels, which results in being better, faster, and cheaper on all shop floors of a product producing company as well as its supplier and contractors.

<u>Offshore Manufacturing and Joint Ventures.</u> For lack of resources, companies form joint ventures with other companies to develop and manufacture products. For example, General Electric and Pratt Whitney joined forces to introduce a new line of engines to the aircraft market. Other companies decide on strategic alliances to fend off overwhelming competition. Formalizing cooperation between two or more partners provides manufacturing companies with an opportunity to combine engineering and manufacturing capabilities.

Domestic manufacturers, burdened with high labor costs and logistics problems sometimes elect to have their products made in other countries. Often this is done to bring them closer to the market they want to sell to and service. A case in point is the automotive industry — with the trend toward so-called world cars — which utilizes component manufacturers that span the globe. Manufacturing in concert with other plants spread over several continents complicates an already complex scenario even further. The success of a manufacturing company with such widespread, interlinked production facilities is dependent upon disciplined, well orchestrated manufacturing principles, goals, and performance measures.

Technical Complexity. Adopting a myriad of new, innovative technologies and synthesizing them is the name of the game for world-class manufacturing companies. The complexity of the technical process on the production floor is always-increasing. Exotic workpiece materials — for example, titanium-based alloys — or the machining of parts with dissimilar materials presents a challenge to finish the parts economically. The shrinkage of workpieces, which designers put ever more functionality into, has also pushed machining processes to their limits.

The speed of the modern machining environment puts a strain on every imaginable production resource. Combined with the need for greater productivity, predictable high quality, and automated monitoring and inspection, it becomes clear that the degree of technical complexity and the demand for knowledge and expertise is high. Only a pool of technical and technological experts can cope with the associated manufacturing problems.

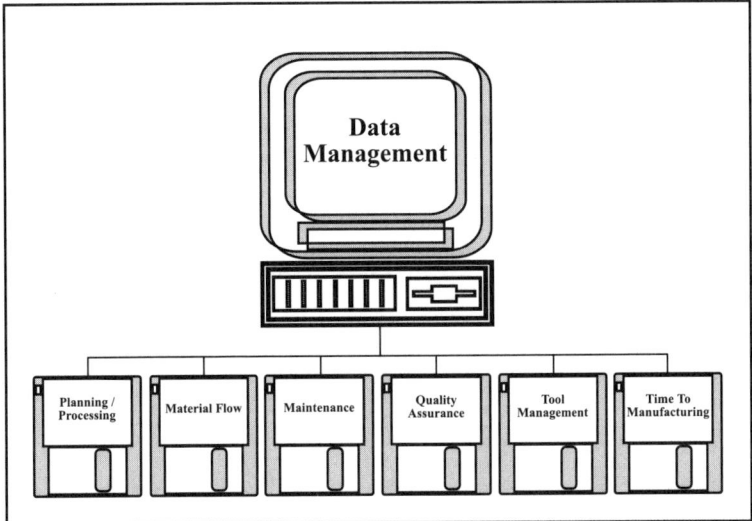

Figure 4.11 • Data Management

Data Management. Accurate information gathering and efficient communications are essential to synchronize the manufacture of products through multiple plant locations. Every plant should have a shop floor control system, with a database consisting of the shop floor functions illustrated in Figure 4.11. The compilation of the data bank is essential to the concurrent manufacturing team.

Analyzing, planning, and processing data can indicate whether production times, for example, are safely used, or whether a process uses too many tool changes or too many machining operations. The smooth and uninterrupted flow of materials (workpieces and support parts) can provide the team with information about both internal and external supply lines, particularly in conjunction with just-in-time (JIT) deliveries. Other parameters would be to optimize material compositions for machinability or other traits that influence production processes.

Failure of tools and machinery (despite preventive maintenance programs) can be systematically analyzed to determine the cause, and process improvements can be implemented with defined maintenance data. To the concurrent manufacturing team, the database can impact quality assurance as it provides a monitoring system delivering the most relevant messages for continuous improvement. Machine capability studies, FMEA's (failure mode analysis), cutting tool life, and SPC as a whole, tell the team where to interfere and innovate for corrections and enhancements.

Applied tool management systems collect all relevant data from the start to the end of the machining and manufacturing process, including tool selection, setting, handling, monitoring, and, above all, feedback of the process with respect to the results at the end of the "line." This feedback is comprised of such essential information as geometric part finishes, tool wear, and tool life, as well as recommended adjustments to correct problems. The aspect of machining and manufacturing times (time to manufacturing) deals with the duration of the entire process and the individual times in between. These are important indicators of productivity, and serve as yardsticks relative to response time, product mixes, erratic lot sizes, and the plant's capacity.

Of course, data from planning, processing, materials, material flow, maintenance, quality assurance, tool management, and time to manufacture are all used for specific and absolute cost comparisons.

Communication. The transmission and exchange of data from the plants to the concurrent manufacturing team (and vice versa) can be through formal and informal vehicles as well as electronically or by fax, telephone, memos, and so on. It is important that the information shared is timely, relevant, and accurate. Communication must occur from the top down, horizontally, and bottom up, so that all levels of the organization are informed. The team should receive relevant information firsthand, and then see that requirements and recommendations are relayed directly to where they can be used to improve the process.

GLOBAL MANUFACTURING

In a shrinking world, there is the chance to reap the benefits of globalized strategies. Going around the world to purchase, design, build, and market products internationally and jointly offers immense opportunities to every manufacturing company.

Strategic Considerations

Today's customers want innovative, high quality products "for the lowest price," and they want them immediately. If company "X" cannot provide a superior, technologically advanced product, then company "Y" will.

In an impatient market, customers demand action, competitors force action, and technology provides action. When a manufacturing company decides to go global, it does so for growth reasons: to sell more by expanding to new markets, to be a local player, to provide enhanced service and responsiveness to local needs, to be more familiar with local environmental and health issues, to be strategically located also for human resources, and to save transportation costs. These benefits will in turn provide a larger return on investment, concessions from local governments (subsidies, protection from other imports, enticements to export to other regional countries), and hedging against political instability of foreign governments and exchange rate fluctuations.

Cable and satellite information and communications reach the entire world community, bringing customers and competitors closer together into a global village. Technological advances can quickly become industry standards until replaced by yet newer advances and technologies. Strategies for global manufacturing have to include the fact that customers now reach for (and demand) the impossible, and have all but abandoned brand loyalty. Globalization's gravity is technology, around which more and more customers and competitors orbit. Manufacturing companies can no longer ride out the traditional "S"-shaped product maturity curves. Long before a product reaches the "S"-curve plateau, new technology often eliminates the old one, since all manufacturers try to ride the same curve.

Many of today's products are of immense technical intricacy, composed of multiple parts and components made or assembled by different plants, often in various countries. Globalizing the manufacturing processes involved invariably challenges one's own capabilities and opens up the quest for new technology. A global organization has a need for worldwide knowledge transferred into all directions, as well as a need for learning.

David Garvin has stated that: "A learning organization is an organization skilled at creating, acquiring, and transferring knowledge and at modifying its behavior to reflect new knowledge and insights. This must be part of any global manufacturing strategy."

Managing Global Manufacturing

For technology driven companies, global manufacturing does not necessarily mean to produce identical products on several continents. It can mean to produce similar parts in geographically diverse areas of the world, or totally different parts for the assembly of the very same product or an end product with regional nuances to satisfy local tastes or regulations. A typical example would be the so called "world car," which may be typically dispersed or network manufactured in the following manner.

- Identical engines and transmissions are made in several countries around the globe.
- Steering parts are made in America and in one country in Europe.
- The same steering parts, because of capacity limits, are made by subcontractors anywhere.
- Vital support parts for transmissions, engines, and steering systems (such as pistons, valve bodies, and cylinders) are made by other subcontractors dispersed around the world.
- Engines, transmissions, and steering systems, together with other subassemblies produced in several plants, are eventually shipped to one or more plants for final vehicle assembly.

Global manufacturing means producing (almost) identical parts, components, and subassemblies with the same quality and within the same time frame and comparable cost structures. Managing and coordinating them successfully is done through concurrent manufacturing by using a multinational, multifunctional "think-tank" to identify and pass on and make mandatory the use of the best process available for producing the product.

What processes should be made uniform, and to what degree, and how much room should the team leave for local deviations from the "norm?" A car manufacturer, for example, can only accept parts with the same geometric finishes. The concurrent manufacturing team will recommend the best suited machines and tooling. This can mean either using the identical machining and manufacturing processes for all tooling makers, or giving the individual plants a window to buy local tooling to assure the same identical process and manufactured results. See Figure 4.12.

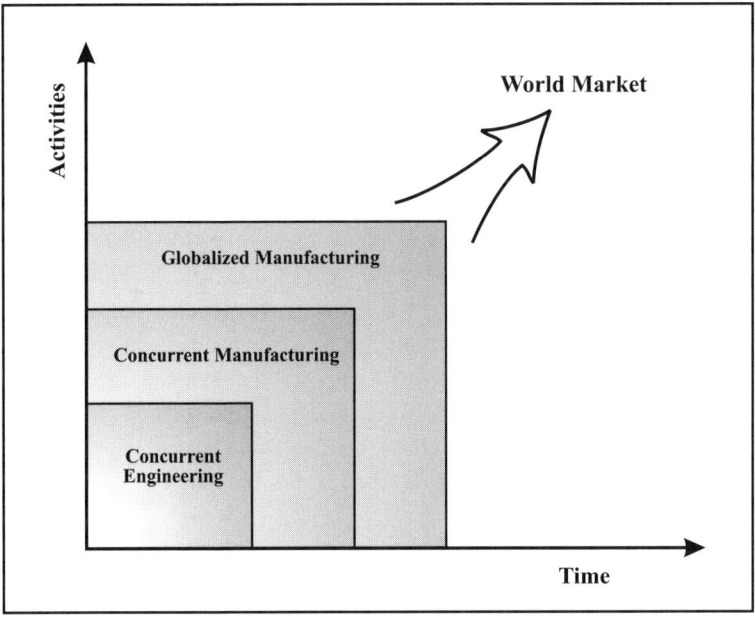

Figure 4.12 • The Development Toward Globalized Manufacturing

A degree of flexibility for individual plants is necessary, and the plants should be allowed to pursue certain manufacturing and technology support activities locally. These activities include:
• research and development,
• certain procurements,
• day to day planning,
• production engineering,
• quality management,
• tier 1 or tier 2 suppliers, and
• product and design engineering for local adjustments.

Commonality of processes leads to homogeneity of products. Commonality, however, must not preclude room for innovation. Headquarters and the concurrent manufacturing team must carefully weigh the extent to which processes are to be standardized. Again, the objective is to produce interchangeable parts at the same quality level, with cost and time being the common denominators. In fact, the impetus for new technology and processes and the subsequent diffusion often originates in "sister" and subcontractor plants.

Another aspect of partially deviating from the best process recommended by the team would be a plant's capacity or higher than anticipated production volumes. The allocation of dedicated machinery as opposed to a flexible production environment, either semi or fully automated, can have a bearing on the decision to deviate. Consequently, the opportunity must be provided for the individual plants to procure local technical and technological adjustments and enhancements as necessary. However, globalizing manufacturing (processes) has to be done in a methodical and disciplined fashion.

- Consolidate the pool of knowledge available, arrive at the desired quality level early on in the development stage, eliminate duplication of effort, share research and development, select the best problem-solving methods, and provide processes that achieve proven and predictable results.
- The development and subsequent application of manufacturing/ machining processes by one manufacturer for worldwide usage preempts the manufacturing environment of individual countries. The key is to design and develop manufacturing processes in a combined effort, in such a way that they can be transferred and copied at any production facility all over the world, or can easily be modified and adjusted if local considerations so dictate.

It stands to reason to recruit suppliers for common goals in order to jointly develop engineering and manufacturing processes. Concurrent manufacturing is the preferred medium for coordinating the activities, and harvesting substantial benefits for the whole organization, its contractors, and selective suppliers. See Figure 4.13.

World-Class Manufacturing

Worldwide manufacturing doesn't necessarily mean world-class manufacturing. (See Figure 4.14.) And manufacturers with aspirations of being worldwide can't equate this with being world-class. Defining "world-class" manufacturing encompasses the benchmarks of cost, quality, and time, plus continuous improvement. These are also the elements of competitiveness and customer satisfaction, and the enablers of success anywhere — locally, domestically, or internationally.

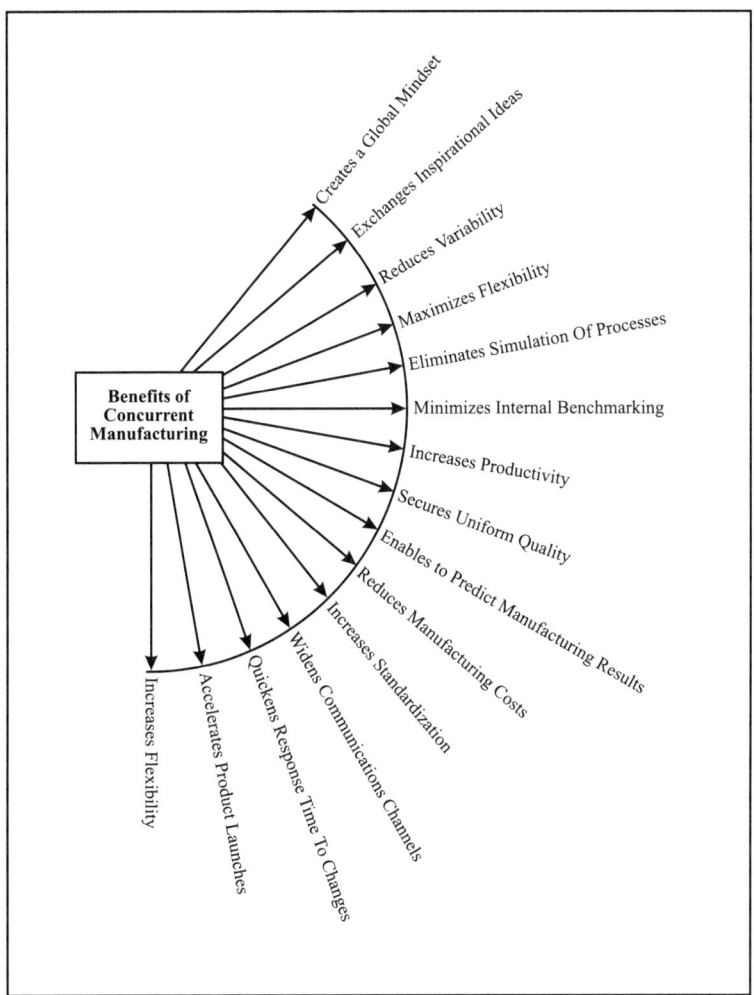

Figure 4.13 • Benefits of Concurrent Manufacturing

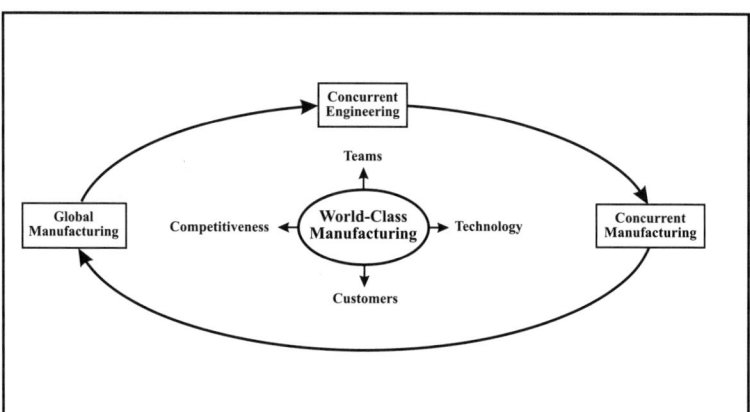

Figure 4.14 • World-Class Manufacturing

Applying and exhibiting the best available practices through teamwork and technology strengthens the company's internal competitiveness and its external ranking in customer satisfaction. These are the benefits of placing concurrent manufacturing, evolved from concurrent engineering, at the core of the manufacturing process. With this, global manufacturing can indeed be world-class manufacturing.

Chapter
5

No-Nonsense Manufacturing

Introduction

In the 1980's, M.I.T. researchers conducted a study of the U.S. auto industry's worldwide competition — most notably the Japanese. Their benchmarking for performance in the industry revealed that U.S. manufacturers still pursued the old production principles of "mass production," while Toyota's worldwide success was based on the principle of "lean production." The book *The Machine That Changed the World*[1] provided research-based evidence that the method of lean production yields superior results in cost, quality, and time. Ever since then, American and European car manufacturers have embraced lean production, and have undertaken massive changes in their manufacturing methods. Lean production involves Total Quality Management (TQM), flexibility, teamwork, volume (and fast) throughput, short delivery times, and design for manufacturability in order to eliminate waste and instill continuous improvement.

U.S. manufacturers now need to concentrate on how to further improve and combine the lessons learned through the original

• • • • • •

[1] Womack, James, Daniel T. Jones, and Daniel Roos. *The Machine That Changed the World*, New York: Rawson Associates, 1990.

principles. One topic we need to revisit is the issue of supplier involvement. The principle of agility within manufacturing also deserves more attention, and there is still much more room for simplicity and realism on the production floor. Although the adage of "organization before automation" is widely recognized, it is rarely pursued, and today's management matrix is still not commensurate with the complexity of today's manufacturing environment.

If we can get better control of these areas, we will have taken a giant step toward sensible manufacturing.

SUPPLY CHAIN EFFECTIVENESS — THE COMPETITIVE EDGE

According to a survey by the University of Michigan, 50% to 75% of the production of manufactured goods in the U.S. is out-sourced.[2] A study by the Arizona State University indicates annual declines in the number of suppliers of about 5%.[3]

The above statistics have interesting implications. First, the extraordinarily high percentage of products outsourced suggests that original manufacturers are heavily dependent on their suppliers. Second, the shrinking number of suppliers points to consolidation efforts in the customer-supplier relationship. What is the meaning of these statistics? Is there an equilibrium in the manufacturer-supplier relationship? Is this relationship beneficial for manufacturing?

The answers are found when we examine the reasons for the fundamental changes in the traditional supply chain. When major industries were forced to slim down and become more productive and efficient, they examined ways of cutting costs. They found that making purely economical decisions on whether to manufacture a product or purchase it from a supplier — as long as the delivery time integrity and product quality were not compromised — was most effective and profitable. Now, more and more manufacturers reach out to third parties for finished products. There are several reasons. One driving force is the cost of the acquisition of capital equipment. The efficient production of new generation parts might call for new machining systems. The original manufacturer might not see the need to invest in new, expensive, dedicated

· · · · · ·

[2]Status, 1995
[3]Status, 1995

machinery in order to produce the part internally. Second, the OEM then would also not be concerned with personnel requirements — another aspect of the make or buy decision: Qualified production people are more difficult to come by, given the dwindling skilled labor pool; and by limiting production activities, larger manufacturers have less need for qualified talent.

Considering the speed at which technology changes, and the velocity with which today's technology is replaced by new innovation, it simply is too difficult for big manufacturers such as Boeing or General Motors to be up to date in all areas of their manufacture. It makes more sense for them to look inward and concentrate on core competencies. Doing R&D, product and process development, engineering, *and* manufacturing for secondary or auxiliary products is just not practical. Big manufacturing has come to realize that one company cannot be best in everything, and better quality might indeed be manufactured by product expert companies. A third important driver of relying on suppliers is inventory, which is related to cost and timely availability of goods and materials needed for finishing products. Rather than produce and stock parts and components in huge amounts and make them available as required, companies now rely on the timely supply of such goods from the outside as they are needed, and not sooner. Just-in-time delivery has become an outstanding tool for cost reduction.

CRITERIA FOR CHOOSING A SUPPLIER

The criteria for choosing a supplier are quality, delivery, and price (see Figure 5.1). Depending on the state of the economy or what the backlog of the manufacturer looks like, the priority ranking of the three basic categories might change. In the wake of quality awareness and standardized policies and procedures, e.g., ISO 9000, the quality of products has become almost a given. The purchase price for a product when order books are full is not as important as delivery time and delivery integrity.

Supplier Types and the Advanced Flow of Suppliers

Depending on how they fill the respective need, there are different players in today's advanced supply chain. (See Figure 5.2.)

Tier 1 Suppliers: When the automotive industry narrowed down its supplier base, it began to rank suppliers not only on their ability, but also on importance, frequency, and value of their business to the

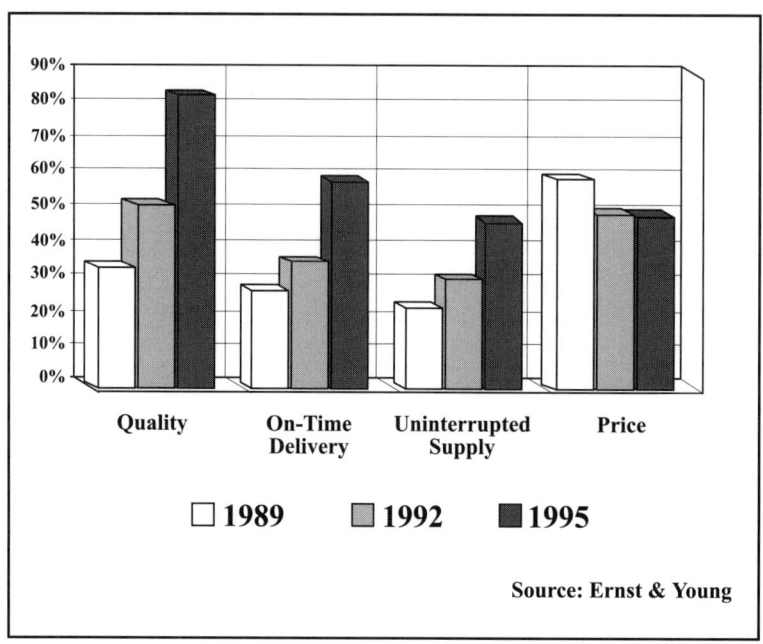

Figure 5.1 • Prime Criteria for Choosing Suppliers

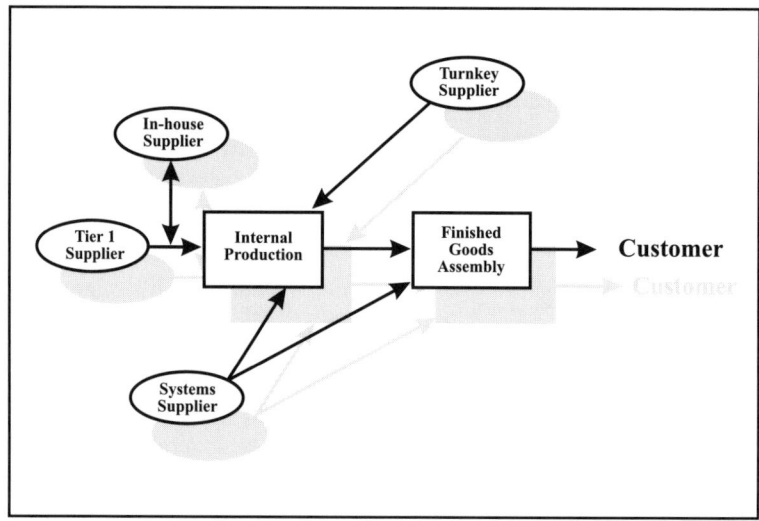

Figure 5.2 • Main Types of Suppliers

manufacturer. A Tier 1 supplier typically manages an entire area of responsibilities, for example, the purchase, inventory, usage, and sorting of tools for one or more automotive plants. The Tier 1 supplier usually enters into contractual agreement with the manufacturer and in turn can select the respective subsuppliers.

In-House Suppliers (Representative): Just as with the Tier 1 supplier arrangement, the in-house representatives of a supplier (often Tier 1), are experts on their company's processes, parts, and materials — both technically and commercially. They impart that knowledge to the manufacturer's floor personnel and are involved in their day-to-day business. This arrangement allows for first-hand planning and rapid response.

Systems Suppliers: A manufacturing company that makes complete subassemblies, for example, power steering assemblies for cars, is considered a systems supplier. Their emphasis is on both high quality manufacture and JIT delivery and inventory control.

Turnkey Supplier: When a car company decides to build a new automatic transmission, it selects a machine tool builder to manufacture and install a new manufacturing system, including processing, machinery, tooling, fixturing, monitoring, transportation, etc. This is called a turnkey project — a system that is installed and ready to use when it is handed over to the customer.

Challenges and Opportunities for Suppliers

As the advanced supply chain opens up new avenues for manufacturers and their preferred suppliers, what are the challenges and opportunities for the supplier that enters into an agreement with an almost always larger and more potent manufacturing company? The main challenge faced by preferred suppliers is maintaining, controlling, and shipping inventory *as needed by the manufacturing customer*. For the manufacturer, receiving deliveries as needed and within short time spans also allows for design changes on short notice. Just-in-time delivery has become a major tool for reducing cost. It streamlines production by minimizing inventory quantities necessary for maintaining delivery schedules. The complexities involved in providing inventory are comprehensive, and suppliers have to manage every facet of just-in-time delivery including inventory control, shipping, handling, minimum level stocks, electronic data interchange, inventory reporting, volume fluctuation, etc.

A preferred supplier must be committed to the ongoing process of continuous improvement in product, process, and service in order to benefit themselves, the manufacturing customer, and the end user. Since most suppliers use subsuppliers, they have to carefully select them and apply the same performance yardsticks for cost, quality, and time. Preferred suppliers are regarded as experts in their field, and manufacturing companies depend on their knowledge and expertise regarding the product and process. (See Figure 5.3.)

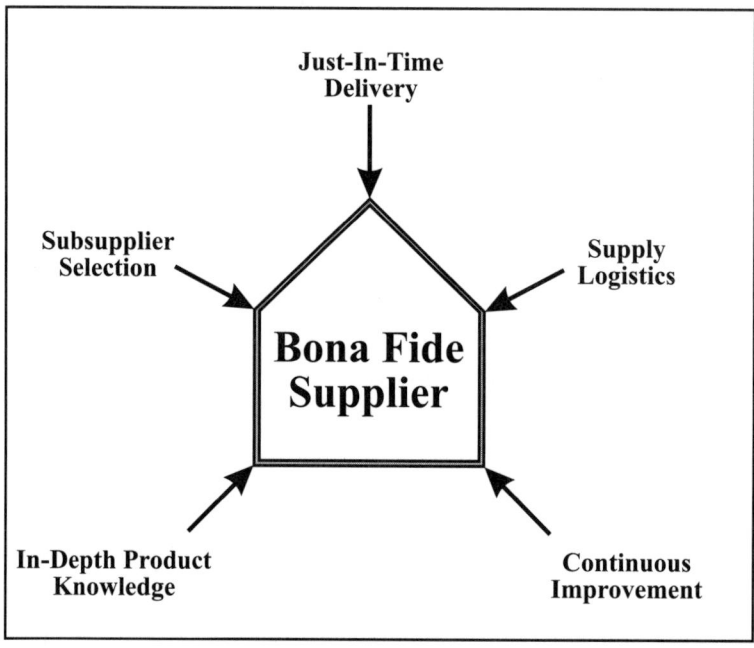

Figure 5.3 • Demands on a Bona Fide Supplier

Becoming a preferred supplier for a larger manufacturing company opens up new opportunities. Concentrating on certain products is a chance to specialize and become a product leader, and this can lead to supplier agreements with other customers. Growing from a preferred supplier into a preferred *systems* supplier offers more leverage and added value because the applied knowledge can translate into other product lines or systems. Handling the logistics end of the business — for example, tool management, or inventory control, or product planning — provides a wealth of information for the supplier to use as an advantage over competition. A preferred supplier who enters into a contractual agreement with manufacturing customers assures itself of business over a certain time period. This frees the company, as shown in Figure 5.4, from having to hustle for orders, and allows it to concentrate on the business at hand as well as its own core competency.

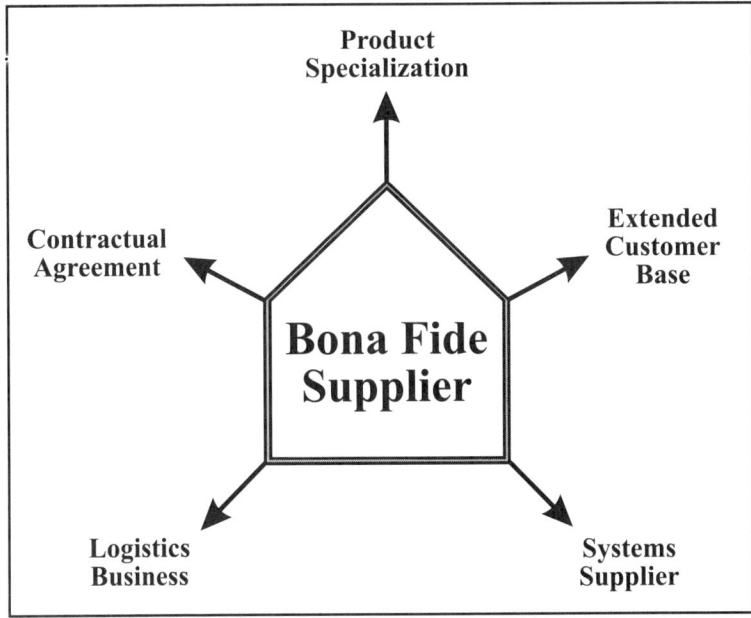

Figure 5.4 • Opportunities for Bona Fide Suppliers

Pitfalls of a Customer-Supplier Relationship

It is certain that the days of the "good old boy" relationship between customer and supplier are gone — advanced supply chains have replaced the ineffective, expensive, and unproductive outsourcing of parts and material. Today's paradigm of quality, delivery, and price puts extreme pressure on manufacturers and their suppliers. There is a lot at stake for both, but difficulties threaten much needed progress.

The Supplier — In a Quandary

The selection process for preferred or certified suppliers is rather vigorous, and the demands placed on them may be difficult to meet. While it makes perfect sense to conform to ISO 9000 requirements and to adopt JIT and kanban techniques, suppliers are hard-pressed to install customer-specific software for every individual account. This holds true particularly for small- to medium-sized suppliers who often don't have the financial, technological, and logistical resources available to them.

Manufacturers expect the following from bona fide suppliers (see Figure 5.5):

- the right price,
- absolute delivery integrity,
- innovative products,
- just-in-time delivery,
- consistent quality levels,
- overall cost savings, and
- continuous improvement.

---

1. The supplier knows the management philosophy of the purchaser and continuously and actively maintains contact with the purchaser. He or she is also cooperative.

2. The supplier has a stable management system that is well respected by others.

3. The supplier maintains high technical standards and has the capability of dealing with future technologies.

4. The supplier can supply precisely those raw materials and parts required by the purchaser, and these meet the latter's quality specifications. The supplier also possesses process capabilities for that purpose or has the ability to enhance such process capabilities.

5. The supplier has the ability to control the amount of production or has the ability to invest in such a way to ensure its ability to meet the amount of production.

6. There is no danger of the supplier breaching corporate secrets.

7. The price is right and the date of delivery is met precisely. In addition, the supplier is easily accessible in terms of transportation and communication.

8. The supplier is sincere in implementing contract provisions.

Figure 5.5 • The Characteristics of Bona Fide Suppliers

The expenditures for capital investment and the organizational stress might well be beyond the pain threshold of some enterprises. But even if they can make the commitment, the next problem they face is that they can often only be a supplier for one or perhaps a few manufacturers, usually in the same industry. That makes them vulnerable because they go through the same market peaks and valleys as the manufacturer, and eventually may be out of business if they do not have the financial strength to sustain economic downsizing. A built-in danger for suppliers is investing in the machinery and workforce necessary, only to have no funds left for research and development to secure new innovation. Over time, the supplier will be left behind technologically, and discarded.

The principle of "single sourcing" has resulted in many suppliers looking for mergers or acquisitions in order to be the one-stop supply house that big manufacturers often look for. Single source responsibility, however, can mean such a full plate of different products that core competencies might be lost over time and the diversification may result in providing a greater array of products. The other trend in the automotive and aerospace industry is known as "modular sourcing." This means that the manufacturer only wants to do business directly with component or systems suppliers. Any supplier depends on subsuppliers (category of Tier 2, Tier 3, and so forth) as part of the supply chain; so the first supplier has to pass the same or similar demands of cost, quality, time, support, and knowledge down the line to Tier 2, Tier 3, and below. In the majority of cases, these demands go well beyond the ability and capabilities of the extended supply chains to the detriment of otherwise capable suppliers. And, as if this was not enough, the newest element of "global sourcing" adds more volume to a seemingly saturated pool of suppliers.

The Manufacturer — Self-Induced Risks

Having forged the advanced supply chain — which has, to a great extent, been instrumental in their competitiveness — manufacturing enterprises find themselves in the midst of a major dilemma, and they are faced with risks they could not have anticipated:
- imminent techno-drain,
- not enough involvement,
- too much dependency,
- smaller pool of knowledge,
- break in the supply chain,

- too high expectations, and
- push to save at any cost.

Giving suppliers too much decision making control can preempt a manufacturer's own ability to deal with day-to-day improvements, long-range upgrades, and up-to-date information on new techniques and processes. Draining one's own manufacturing and engineering staff purely for cost-cutting reasons can be a perilous proposition because the involvement and competency of skilled employees is needed. How can a manufacturer give an order for a complex machining system to one supplier for a turnkey project, knowing that their expert knowledge is only in one field (let's say machine tools, but to a lesser extent cutting tools), and that the end user is only willing to take the product if the initial part runoffs of the machining system are satisfactory. Due to this, manufacturers often find it necessary to correct defective purchases. The installation of in-house management by a supplier not only not adds value to the manufacturing process, it also further reduces or even eliminates other outside sources. In other words, it creates a scenario where healthy competition is a foregone conclusion.

This leads us to the concept of single source responsibility — the selection of a single source to supply total production of a part. Historically, manufacturers chose several suppliers to keep costs down and avoid dependency on one supplier. Single sourcing, with all its associated cost savings, needs to be improved upon. When General Motors output came to a screeching halt due to a strike at a single plant that produced a key car component, they couldn't meet pent-up demand for several car models, and so lost precious marketshare. The combination of single sourcing and just-in-time delivery can be especially devastating, and not only in terms of not having parts available "on the line." It can strain the manufacturer-supplier relationship to the breaking point in the event of unpredictable sudden backlashes.

As it turns out, awarding a contract to a Tier 1 preferred supplier and then letting them manage the second, third, and fourth tiers, is near-sighted. The down-the-line suppliers are faced with cost-cutting demands, have to outbid competitors to get awarded projects, and are used as buffer stocks for the ups and downs of market demands. All this may be unknown to the manufacturing enterprise. It must be remembered that a break in an established supply chain is always a possibility, and that even the nominally least valuable part can interrupt production of a large complex end product. Clearly, today's supply chain techniques, and the sensitive and vulnerable interrelation of manufacturer and supplier, needs improvement. The benchmarks of cost, quality, time and knowledge need to be revisited and adjusted.

Making It Work – The Advanced Supply Chain and World-Class Competitiveness

The Supplier — Room For Improvement

A survey conducted by the Industrial Technology Institute, Ann Arbor, Michigan, revealed that suppliers to the automotive industry (employing up to 500 employees) in the mid-1990's are less productive than they were in the 1960's, and only a few have soaring productivity.[4] For suppliers to be considered a part of an advanced supply chain, it is imperative that they invest in modern capital equipment and skilled labor. The key is to be flexible and fast in response, and to have the capability and ability to extend knowledge and become a systems supplier — that is, to be a company that can provide functional subassemblies as well as discrete parts. A supplier must offer technical and technological knowledge, and must be ready and willing to communicate with the original equipment manufacturer on design and process. Electronic data transfer link with the manufacturer gives the supplier insight into what technology can be anticipated, or where to acquire it if it can't provide it on its own.

Communicating with subsuppliers for cost, quality, time, and knowledge is an important part of the supply network. What works for the OEM also works for the Tier 1 supplier — the quality must be there when a part or assembly is received from lower tiers. There are hardly any stand-alone enterprises which operate entirely independently. One of the elements of being an attractive supplier is to apply the principles of continuous improvement. The customer needs to believe that the supplier is commercially, technically, and technologically competitive. An advanced manufacturer-supplier relationship can only be built on mutual trust. The supplier can do its part by openly sharing its limitations as well as new innovation, as long as its counterpart is willing to reciprocate essential information in confidence. This is a prerequisite for developing into a possible partnership agreement.

• • • • • •

[4]Automation News, December 1996

Especially among the top-tier suppliers, there is an urge for acquisition and mergers to expand marketshare and become more competitive in a global environment. The smaller supplier will have to follow where the top tiers go and where the OEM's want to be. Suppliers have to try to ramp-up for geographically distant locations, including going overseas, to stay as close to their customers as possible. For suppliers to succeed in the new supply chain, they must be open to new opportunities, and be willing and able to adjust to change.

What the Manufacturer Must do Better

In order to develop a pool of reliable suppliers, the OEM's will have to be knowledgeable and skilled in the areas of finance, communication, and in building trust. They will also have to realize that keeping competitors available is prudent, and that inventory issues will have to be addressed. Being a "good" customer has its advantages, too.

For the smaller supplier, the financial burden associated with accommodating the manufacturer's demands can be too much to handle. Any assistance toward long-term inventory or the purchase of expensive equipment can make a difference for suppliers to deliver what, when, where, and how (in what configuration) the customer wants. The JIT concept can be the Achilles heel of lean manufacturing, and can quickly disrupt the supply chain.

The pull concept of inventory is to reflect actual demand, which is usually short- term and not medium- to long-term forecasting. Manufacturers need to reconsider some inventory strategies. With dependency on one or a few suppliers, and with the supply pipelines getting longer through global sourcing, it is mandatory to keep a supply of critical parts in one's own stock or at a vendor close by. The cost for maintaining some inventory is not as high as the loss of customers for lack of a product. Regarding global supply lines, it would behoove the OEM to ask their technology suppliers to move with them offshore. From a practical standpoint, it involves relocating, but it assures the continuation of "well-established relationships" that would be beneficial to both enterprises.

Unrealistic requirements such as short term double-digit price cuts or quality part finishes that the manufacturer cannot sensibly achieve itself should not be part of the stipulations set forth to the supplier. It makes good business sense to have competitors available, particularly for critical components. Global sourcing can assure both cost and quality. Single sourcing has its merits when the manufacturer needs to slash costs. Again, it is imperative to keep alternative supply lines open.

Working principles of the advanced supply chain are set in motion by the manufacturer. Most suppliers feel that they have to take what is imposed on them, for better or worse. Instilling trust is one of the prime tasks for any manufacturer in terms of business fairness and loyalty. The manufacturer-supplier relationship is a two-way street, benefiting both parties.

Supplier Integration — At the Heart of World-Class Competition

Manufacturers and suppliers who feel confident about each other are viewing partnerships as a practical means of extending their capabilities and discovering new horizons. Rather than creating antagonistic relationships within a complex supply chain, it is wise to abandon the traditional arms-length process of buying and selling, and create a structure based on cooperation and integration. (See Figure 5.6.) Two or more companies working together can obtain more benefits than stand-alone enterprises operating on their own. Supplier integration to the extent of partnering is an important link in the advanced supply chain. Good internal and external relationships based on trust and open information sharing are the prerequisites of a healthy partnership.

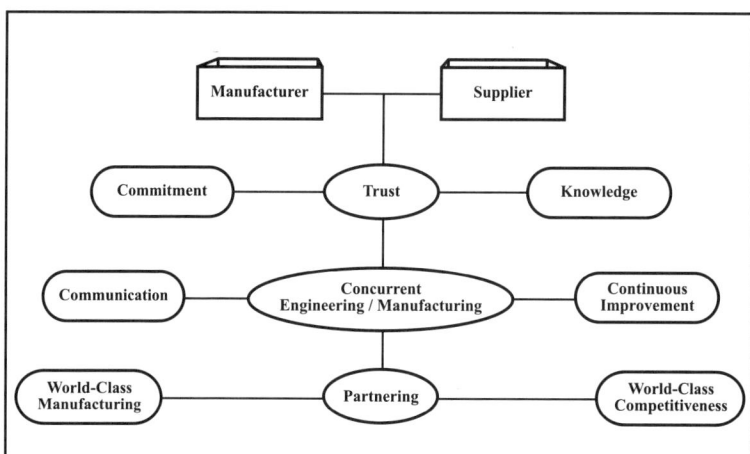

Figure 5.6 • Manufacturer-Supplier Involvement

A contractual agreement stipulates that the partnership is a genuine two-way street that goes beyond the mere supply and acquisition of the desired product, to a relationship with genuine trust and commitment on both sides.

1. Sharing pertinent information about market gyrations or anticipated shifts in end user tastes, or predicting strategies by competitors, can cut lead times considerably. The associated cost savings can be substantial.
2. If the principles of concurrent engineering and concurrent manufacturing are applied in earnest, the exchange of ideas can lead to more frequent innovations in product and process. The open and accurate sharing of information can also ensure a better chance of success for newly implemented projects.
3. As the partnership grows and trust is established, an open cost and price structure helps identify where both entities stand with respect to the product economics, so that necessary adjustments can be made much more easily.
4. The involvement in each other's process and product opens doors for improvements. Having suppliers participate in cost reduction programs and sharing the savings can result in an onslaught of ideas benefiting the manufacturer and supplier alike. As a whole, when both parties are committed to sharing the risks of their business venture, they ought to share the benefits of it, too.
5. The integrated supplier arrangement allows the manufacturer to concentrate all resources on its core competencies. Likewise, it enables the supplier to improve on the effectiveness of their core product.
6. Manufacturers and technology suppliers can share R&D efforts for reductions in cost and development time. This can mean an increase in their R&D competence. Sharing problem solving techniques is especially helpful when manufacturers and suppliers produce the same or similar products (through concurrent engineering/manufacturing).
7. One type of close collaboration is to provide assistance, and that can include lending surplus machinery or skilled personnel in times when one's own capacity can otherwise not absorb order overflows. The mutual benefits are quite obvious.
8. The overriding denominator of the relationship is improved customer satisfaction because supplier and manufacturer measure them-

selves the same way that the customer does. This measurement is based on the belief that cost-cutting is never primary, but the customer is. This collective attitude manifests itself in a product that the customer views to be the most creative and best technical solution — in other words, world-class!

A well-orchestrated, value-added, short supply chain is at the forefront of true world-class manufacturing and competitiveness. It is the yardstick to be measured by in a market where more and more manufacturers of excellence find that partnerships have become critical to long-term prosperity.

STRIVING FOR SIMPLICITY AND CONQUERING COMPLEXITY

Manufacturing, especially the manufacturing of automobiles, airplanes, computers, and many other products is a complex undertaking. But, the manufacture of such products does not necessarily have to be complicated. In fact, if processes and products are too complicated, they are usually too costly to build and difficult to market. To take this further, any manufacturing — including the most complex — can survive only with simplicity in design, process, product, and structure. The benefits of simplified manufacturing can be found in cost, quality, and time.

Webster describes complexity as "suggesting an unavoidable and necessary lack of simplicity, it does not imply a fault or failure in designing and arranging." Indeed, complexity, as such, can be a system, product, or process that is, by its very nature, complex. The way to ward off complexity is to manage it sensibly, productively, and economically. Conquering complexity and promoting simplicity is one of the prime tasks of lean and world-class manufacturing.

Design, Process, Product

It is design that lays the groundwork for all subsequent activities and, to a great extent, influences simplicity for process and product. The principles of Design to Cost, Design for Manufacturability (DFM), and Quality Function Deployment (QFD) have to be applied. Going beyond simplicity in functional and physical characteristics, the cost to produce the designed products as well as one's own manufacturing capabilities have to be taken into consideration. Production should not be given blueprint features or tolerances they are either unable to comply with, or that they can only achieve through extraordinary cost and efforts.

The customer decides on whether the product received meets or even exceeds the expectations in function, cost, and quality. QFD focuses on designing quality and customer needs into the product right from the start, thus reducing engineering changes during the product's life cycle, product development costs, and time to market. QFD is essentially the method of using customer feedback to determine basic design functions.

Simplification in product design will inevitably reduce the number of parts in a product and result in the use of standardized parts and components. The assembly of products is also greatly facilitated through parts made by modular design. Simplifying design is an ongoing proposition to benefit the benchmarks of cost, quality, time for manufacturing, and customer satisfaction.

As for process, there is one major area, other than automation, where there is room for more simplification: specifying the right machining process. The overwhelming combination and varieties of workpieces in configuration and material, cutting geometries, and tooling systems makes it appear difficult to identify the right process for economics, productivity, and precision. However, the state of technology also presents an opportunity for constant improvements. As the major industrialized nations fiercely compete with one another in turning out products their customers want, only the best machining processes are most competitive and determine technological leadership. The optimum machining process is found through the combination of cost/performance ratio, the need for precision, and the machining time. Simpler processes don't have to be less expensive, but they must make machining easier, more predictable, more accurate, more consistent and reliable, and, if possible, faster. If they do that, then invariably they are less costly. The initial tooling cost is, as a rule, higher, because being at the edge of technology has its price; but the price/performance ratio then justifies the purchase (the purchasing branch of corporations has to understand this concept). Engineering has to be able to argue their case for better technology, and Concurrent Engineering/Manufacturing Teams are an excellent platform to communicate for the sake of the best suited machining/manufacturing process, with cost, quality, time, and technology being the determinants.

An example of an optimized machining process is shown in Figure

5.7. This new process involves a universal tool that machines the inner contours of a transmission case. At high speed, it can finish-machine a variety of different bores with minimum cutting force. It is dynamically stable and operates at 4,000 m/min cutting speed at a stock removal of 3 mm. The initial cost for a new tool is about 60% over the cost of the traditional one, but the conventional method of using a multibore or a cutting tool would warp the thin-walled aluminum case.

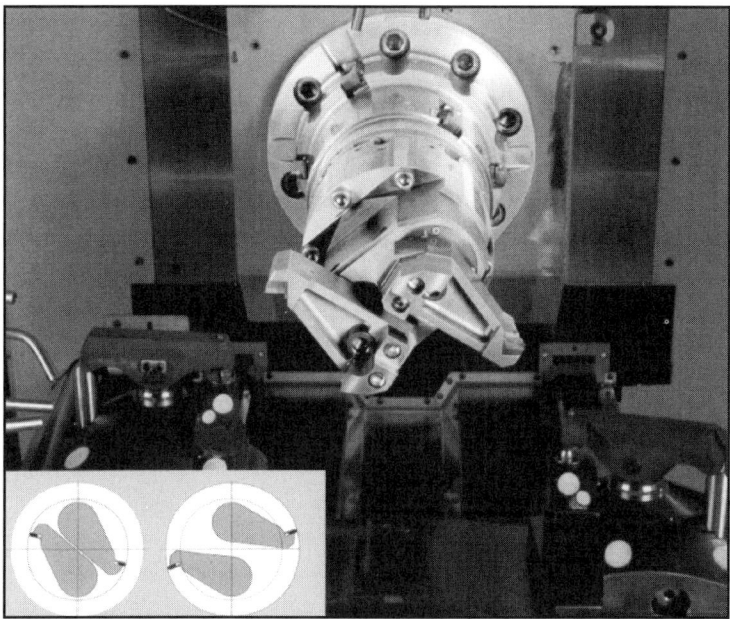

Figure 5.7 • Innovative Generating Tool

The different bore diameters are machined as per the respective centrifugal force induced by the machine spindle's rpm. The tool is adjustable for a certain diameter range and achieves excellent geometric results. The progress toward simplicity of machining processes lies in:
- relatively simple tool design,
- elimination of multibore tool setting,
- higher cutting speed, faster part finish,
- no special part clamping,
- less cutting force, less machine HP, and
- tooling for universal use.

While the multibore tooling configuration previously used facilitated the machining process of using several individual tools, the innovative tooling described here illustrates the ongoing development of new solutions resulting in more simplified production processes.

Automation and Simplicity

The idea that automation makes everything easier and simpler on the production floor is as false as the notion that no automation is less complicated. Automation at any cost could be as wrong as none at all. By far, most parts and part families are made using flexible manufacturing, that is, automation to a high degree — and it is affordable. Although it has not yet fully replaced dedicated machines for mass production, batch production and mass customization have given flexible machining systems an enormous push. It has also become apparent that full-blown flexible manufacturing systems (FMS) are difficult to cost-justify and too complex to keep at a high efficiency rate and uptime. Interlinking a number of machining centers through a host computer, and automatically loading and unloading stations with sophisticated transportation systems for practically unmanned, "lights-out" production can only be realized by large, financially well endowed enterprises.

Most flexible manufacturing is done in an arrangement of a few grouped machining centers that form a flexible production center, called a cell, for a family of similar parts. See Figure 5.8. Those cells provide the benefits of market responsiveness, new levels of consistency in precision, higher machine utilization rates, and computerized production logistics. But the real breakthrough — in terms of simplicity in handling, learning, start-up, and flexibility, plus a more favorable cost structure — is the (almost) off-the-shelf purchase of such capital equipment. Although they remain a complex composition of sophisticated modules, machining center cells made of standardized components have taken a giant leap into the arena of agile manufacturing.

The "plug-and-play" principle is a further development for lean production methods. Lightweight, stand-alone machines can be grouped to cells or simply located next to each other for immediate start-up on any production floor. Part loading, either automatically or manually, adds another element of simplicity. Another important feature is that they are moveable due to their self-contained, built-in, above-ground coolant system and their easy-to-move, lightweight overall construction.

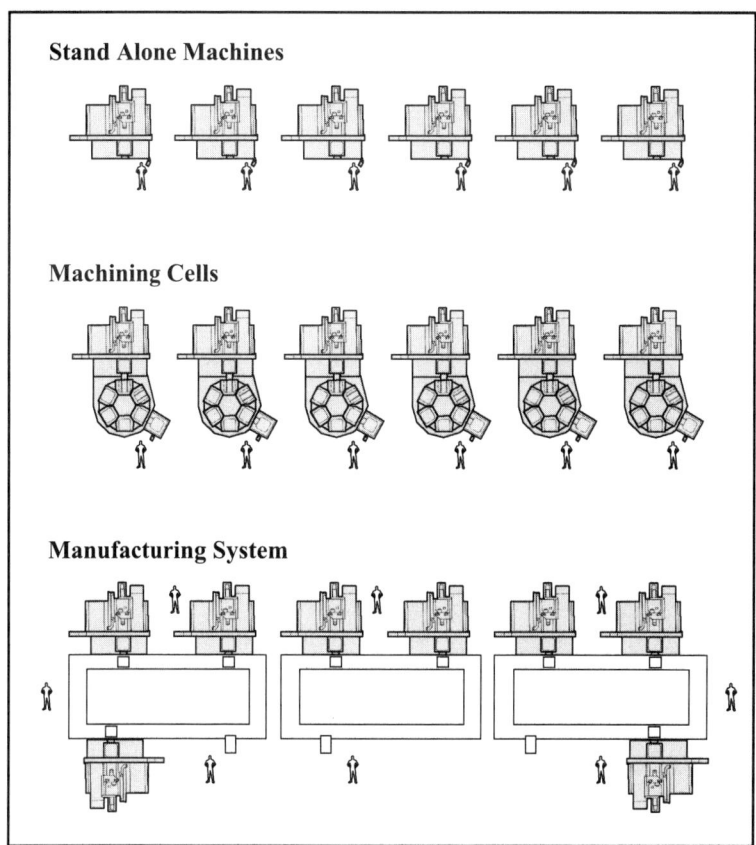

Figure 5.8 • Arrangements of Machining Centers

Consequently, they can be placed into any corner of the production floor to complement other existing machines virtually within hours. By equipping these standardized machines with high spindle speeds and fast axes acceleration rates, as well as extremely low cut to cut times, fewer machines are necessary to complete high batch part numbers. Simple, easy-to-read control systems invite manual operating of the centers, further simplifying processes.

Now machine tool manufacturers need to enhance the simplicity of the process by designing more fool-proof monitoring systems for tool life and preventive maintenance into the machining cell, and by offering uniform software packages. Open architecture is an essential aspect of production floor simplification. Automation and simplicity reach their optimum within flexible manufacturing through standardized, self-contained, high velocity machining centers, which are easily adaptable for different part runs, can be readily loaded/unloaded, and have a high level of immediate start-up capability.

Adaptability and flexibility are the essence of agile, lean manufacturing. However, there is one more element that goes beyond machines and tooling. It is the realistic assessment and realization of the limits of any manufacturing. The hunt for micrometers (mm) has its end somewhere, and so does precision. Specifying unrealistic part finishes that even advanced machining processes can rarely achieve counteracts the quest for more simplicity.

If a certain geometric tolerance can be met without too much stress on processes, we should specify that tolerance. This is above all necessary when affecting the outcome of the end product because the customer wants it to be defect-free. Manufacturing, however, has to be selective in specifying tolerances for critical parts, and must not assign arbitrary, extreme close tolerancing that is neither necessary nor feasible in regular, advanced manufacturing. (See Figure 5.9.)

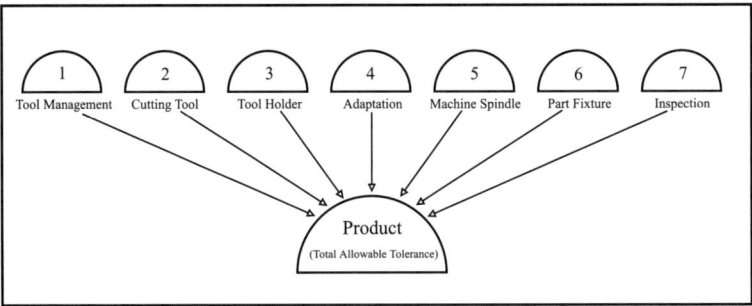

Figure 5.9 • Total Product Tolerance

Part- and Product-Oriented Manufacturing

Flexible manufacturing has eliminated the traditional grouping of machinery by operations. Drilling, milling, tapping, etc., are done on one and the same machining center, often machining complete workpieces. When several stand-alone flex units are grouped together to form a manufacturing cell, they could be used for product-oriented manufacturing or part-oriented manufacturing. (See Figure 5.10.) The decision primarily depends on the nature of the manufacturing company. For a job shop specializing in machining individual parts, part-oriented manufacturing is a given. For a company manufacturing a complete, assembled product, e.g., gearboxes (and only gearboxes) in high quantities, the question then is whether to group cells by part or product. How about an enterprise that manufactures automatic transmissions featuring individual subassemblies? How should the grouping of cells be arranged here? For the sake of production transparency vis-à-vis simplicity, is product- or part-oriented manufacturing better?

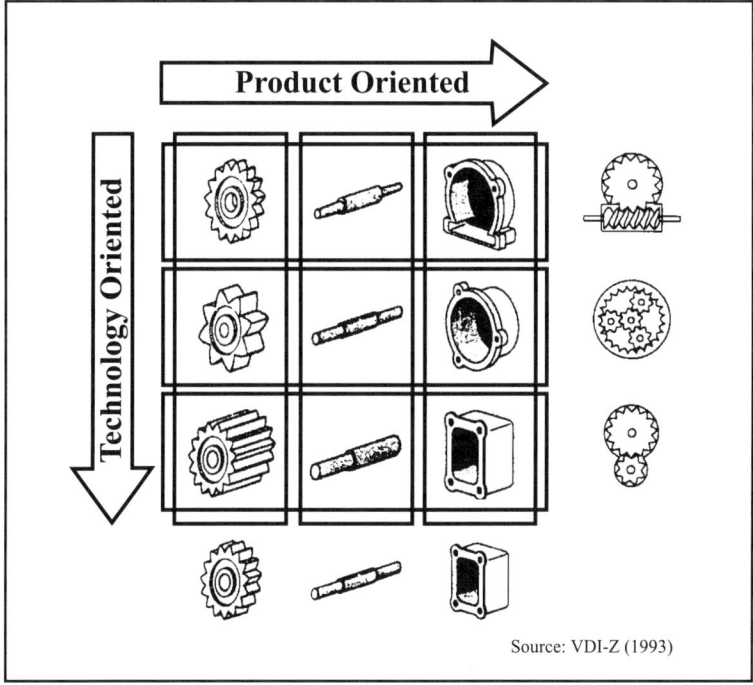

Figure 5.10 • Product- and Technology-Oriented Manufacturing

No Cell Autonomy

The idea to give the manufacturing cell complete autonomy and have it work as an independent profit center is borne out of the desire to commit the operators to increases in throughput and productivity. The concept, however, has major flaws. For an isolated cell operation to be self-sufficient, there is always the issue of capacity within itself, and as it is related to the other production units. Bringing planning and processing to a cell to organizationally make it autonomous creates a company within a company, in isolation. In addition, there is hardly any room for concurrent engineering/manufacturing and continuous improvement principles.

Part-Oriented Cell Operation

A part-oriented manufacturing structure is based on individual workpiece machining, organized by part similarity. Grouping workpieces by similarity saves on part preparation time and setup time, and helps to minimize capital expenditures for part fixturing while increasing throughput time per part. Since other cells produce complementary parts in the same fashion for the final assembly of the end product, there is a need for a somewhat elevated effort in corporate design, planning, and processing. The real advantage of part-oriented manufacturing lies in its adaptability to market demands and cyclical gyrations.

Product-Oriented Manufacturing

Product-oriented manufacturing on the other hand offers complete transparency for all parts because final assembly or subassembly are finished within the cell. Optimizing the factors of cost, quality, and time individually and interrelatedly can be done correctly and easily. Problems in machining and assembly can easily be recognized and, if applicable, be more easily traced back to design. The entire process from R&D to final assembly is lean and transparent. There is another natural degree of simplicity because of a minimum of departmental interfaces. Production of extreme high volume, however, is difficult to achieve.

Part- and product-oriented manufacturing are functions of the product structure of the manufacturing organization. While a product orientation has its limits in machining singular, high-volume production, part orientation can be justified for that reason and as a supportive concept for feeding product-oriented cell manufacturing. Considering the trend toward *small* to *medium* batch production, the JIT principle, as well as the concepts of lean manufacturing and continuous improvement, product-oriented manufacturing is transparent in a single cell manufacturing structure.

Managing Complexity

Innovation pushes technology. Technologically intensive products and processes drive up complexity. In order to be able to produce and market innovative products, it is necessary to manage their complexity. The end product's complexity has to be minimized, and production complexity has to be conquered. Complexity is driven by production, while the organization is driven by product, and a function of all elements and their interconnection within systems. (See Figure 5.11.)

The specific areas of complexity in a manufacturing enterprise are Processes, Planning, Maintenance, Management, Communication/Information, Marketing, and Sales, as well as Product Variety, Functionality, and Service.

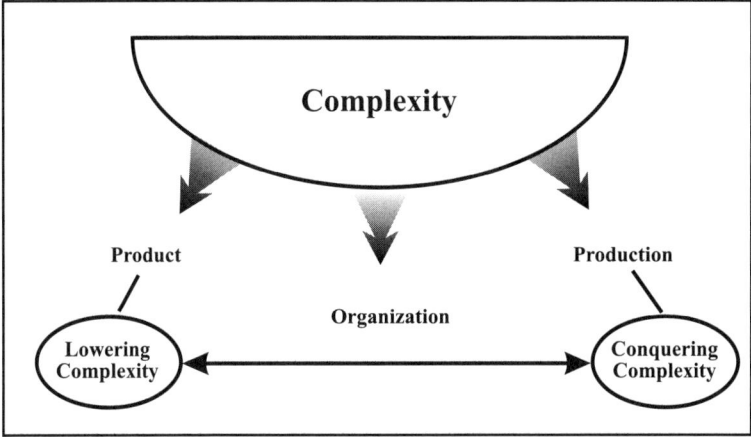

Figure 5.11 • Drivers of Complexity

We cannot afford to let complexity lead us into chaos. Equally, we can not allow it to curb technical or technological progress. In fact, leading industrial nations need innovative, technology-intensive products and processes for worldwide competitiveness and leadership. These products and processes, however, tend to be more and more complex. Their complexity has to be managed sensibly and systematically within certain formats and parameters.

Organization

Management has to strive for minimizing and managing complexity, and it must be structured through thinking in systems, to assure the knowledge of interrelated processes and receive necessary feedback. Since any organization in itself is rather complex (the bigger the enterprise, the

152 • Chapter 5

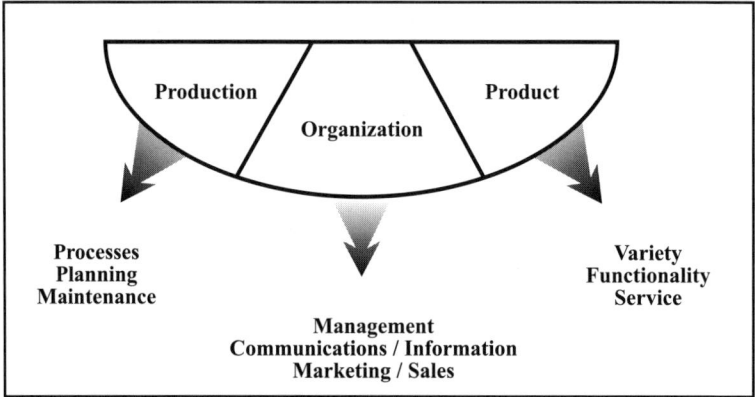

Figure 5.12 • Areas of Complexity

more complex the organization), decentralizing departments or work units and giving them autonomy decentralizes responsibility. Structuring the organization horizontally through modular units opens up transparency. The goal is to cut down on unnecessary departmental interfaces and management layers with a free flow of communication and information organized through working groups such as concurrent engineering/manufacturing teams to distribute information. This foregoes layers of pertinent information, reduces misunderstandings, creates knowledge, reduces risks, and eliminates the "unknown," which is a major contributor to complexity.

The objective of managing the complexity of any enterprise is threefold.

- Structure, systems, and processes of the past have to be revisited and revised, corrected and adapted, to today's requirements.
- Flexibility and adaptability to sudden shifts and changes affecting the organization, internally or externally, have to be exercised with precision and speed.
- Structure, systems, and processes have to be created today to cope with tomorrow's complexity.

These objectives can only be met if the enterprise is willing to learn and has a vision for change.

Product

In order to gain or keep marketshare, manufacturing companies constantly try to differentiate their products from those of competitors. This can lead to increased variety, leapfrogged technology, and tailored product lines. There are several techniques and systematic methods to manage the corresponding degrees of complexity.

Benchmarking: Compare one's own product to the best of the competition, and then design and engineer it to take certain, perhaps unnecessary, complexities out of the product and thereby improve on it.

Concurrent Engineering: Using its principles and concepts right from the start (at the design stage) helps to minimize complexity along with the manufacturing process (since multidisciplinary teams decide on the shape, form, and function of the product).

FMEA: Failure mode and effect analysis is a detailed listing of potential failures in a product. The probability of failure occurrences and their impact on the whole system is explored, and countermeasures are contemplated. This can help prevent potentially serious failures.

QFD

The so-called "house of quality" lists the customer's requests by degree of priority and weighs them against the product features offered. This method shows windows of opportunity to the producer for customer satisfaction of product quality, variety, and acceptance.

Value Engineering

This is a method of solving complex problems. With it, the functions of a product can be reduced to its basics without sacrificing quality and without affecting the customer's demand for product reliability.

Process

With higher degrees of automation, the complexity of manufacturing systems progressively increases because flexible and dedicated machining cells are a composition of mechanical, electric, hydraulic, pneumatic, and software parts and components that need to be tuned to each other in accordance with their corresponding state of technology. Since mechanical parts normally have a longer life cycle than electronics or machine software, upgrading the machines can throw off the originally built-in balance of the system. This can adversely offset service, maintenance, and operation of the machining systems in terms of transparency and parts variety. The compatibility of all parts and components of the machining system is an important part of reducing complexity.

Hardware components of monitoring systems such as probes, sensors, and transmitters have to be commensurate in reliability and accuracy with the software of the system. The whole of the manufacturing process consists of a given number of individual processes which have to be in tune with one another. One weak link in the process can transfer more stress to another in order to compensate for its weakness. Here, too, the compatibility in technology of the individual processes strengthen the process as a whole.

Example: A CNC-machining center capable of running at 18,000 rpm spindle speed and axes acceleration rate of 2 g's has to be equipped with cutting tools designed for high velocity machining and flexibility, as well as tool and part fixturing. Expecting a high chip volume per time unit, provisions for proper chip disposal (coolant through the spindle, coolant volume and pressure) have to be made. If this cell is incorporated within other cells, they all have to feature the same degree of technology for efficiency and productivity. (See Figure 5.13.)

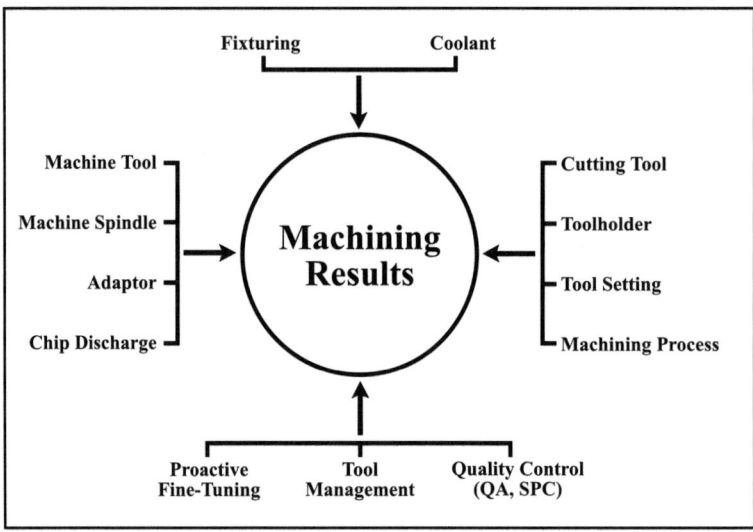

Figure 5.13 • Contributors to Machining Results

It is important to note that the overall efficiency of the manufacturing cell unit must be 85% to 90% in order to justify the addition of a more sophisticated subcell. Keep in mind that a machining and manufacturing system's complexity is greatly reduced through standardization, a minimum of part variety, and the compatibility and transparency of the individual subassemblies.

MEASURES OF ADVANCED MANAGEMENT

The "management by . . ." principles, heralded and adopted in the 1980's, improved the corporate management structure and paved the way for more humanized, logical, and timely corporations. New technologies are rapidly emerging, and customer-expert quick responses in global markets are creating more numerous and formidable competitors. The priorities for management are to establish a culture of cohesiveness; promote creativity and gear up for an agile environment; scale down organizational layers; and assure a short flow of communication. Equally important is the ability to anticipate market gyrations and to adapt to sudden unforeseen changes of the business environment at large.

Lean and Team

The lean concept means more than lean manufacturing. It is lean management, a strategy set forth as a corporation's culture. This strategy embraces old functions of the organization as a process in order to develop, manufacture, and market products of the best possible quality, at minimal cost, in the shortest possible time frame, with the least possible expenditures as shown in Figure 5.14.

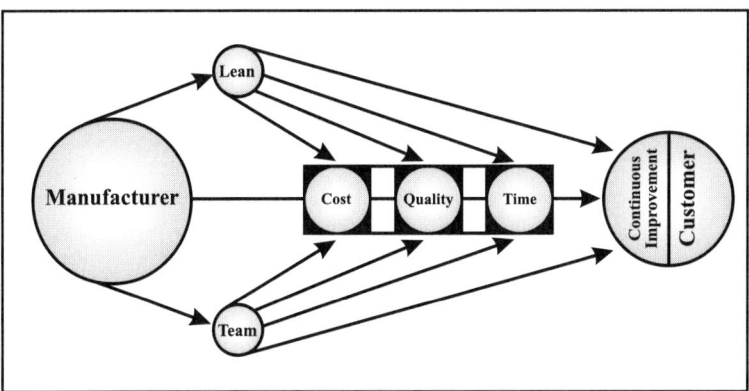

Figure 5.14 • Lean and Team Concept

Lean management breaks with the traditional concept of the Taylor philosophy and the conventional structure of the organization matrix. It can best be understood and implemented — and it functions best — through the coordination of the following elements.

Continuous Improvement of Product and Process: This directly affects the benchmarks of cost, quality, and time, and is best approached through the concept of concurrent engineering and concurrent manufacturing, including the applied principles of "Design for Manufacturing," "Design to Cost," "First Part-Good Part," and "Zero Defect Manufacturing."

Zero Defect: This means absolute, 100% failure- and mistake-free manufacturing. It also should be about striving for failure- and mistake-free work throughout the organization. Because of the intertwined relationship of working groups and interdisciplinary teams, manufacturing can only be done with zero-defect if all other preceding functions and disciplines are performed with the same objective.

Just-in-Time Delivery: JIT is the timely delivery of parts *as they are needed* in manufacturing.

Complete Customer Orientation: The performance measures of a company should focus on the most relevant factor — the customer. Being completely customer oriented, that is, to meet or exceed the customer's expectations, keeps a company in business. Satisfying the customer has to be the goal throughout the organization. It has to be part of everyone's mindset, and must then be acted upon. Only a constant drive for customer satisfaction can assure competitiveness in the marketplace.

Supplier Integration: The relationship between a manufacturer and its suppliers must be commensurate with the importance of the supplier. If the role of the supplier is critical, the relationship must be strong and tight. The integration of suppliers is an important part of the advanced supply chain. Involving critical suppliers in decision-making for product and process can be a decisive advantage.

Advanced Communication: A key part of a lean organization is uninterrupted, accurate, and quick information, communicated informally through memos, circulars, etc. Information and official updates relevant to automatic processing are transmitted through a network architecture such as client-server, host, and PC networks. Given the importance of advanced supplier chains and an ever closer relationship triangle (manufacturer, supplier, and customer), efficient external communication is essential. Being hard-wired in working groups has to be followed by soft-wiring through EDI (Electronic Data Interchange), e-mail, or on-line communication like instant messaging, net meeting, or video conferencing.

By combining internal and external communication networks, a communication structure is realized that spans the entire logistics chain. With its relevant applications, solutions to problems, documents, etc., can be shared and discussed to keep all parties involved up-to-date constantly and immediately.

Organization Before Automation: The desire to increase productivity often ends up in the purchase of automated machinery, and often without regard to other support functions of the existing organization. The essential failure of the CIM (computer integrated manufacturing) concept has proven that fully automated manufacturing on a broad scale is too difficult to manage and support, too costly, and uneconomical. Questionable efficiency rates and the complete dehumanization have turned out to be significant liabilities.

Downsizing and other efforts to improve efficiency throughout the company invariably fail, too, without restructuring the organization. We must accept that there is a practical limit to automation, the degree of which certainly varies from company to company and industry to industry. It is important to realize that a manufacturing company's well-being and future is found first in people and the organization they work in, and second in the machinery. The acquisition of automated machines has to follow the organizational support system.

A More Horizontal Corporation: Simple streamlining and consolidation still leave most corporations with too many layers of management, slow decision making, and high costs. A different organizational model is the horizontal (or flattened) organization. It manages across teams instead of vertically up and down. A more horizontal structure eliminates boundaries between departments and functions, and concentrates on processes run by teams. (See Figure 5.15.)

The key elements of the horizontal organization are:
- core processes determine the company's structure,
- there are a minimum number of hierarchical layers,
- teams manage themselves,
- customer satisfaction is the measure of performance,
- supplier and customer integration is accomplished through open communication, and
- there is efficient flow of information.

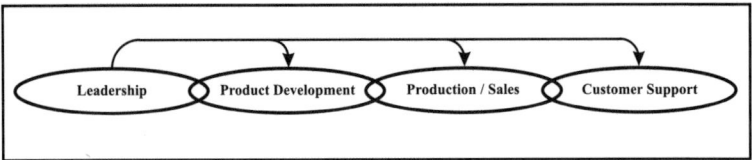

Figure 5.15 • Flat Hierarchy

A horizontal organization in its purest form, with the chairman at the helm and only one or two layers "down the line," can be difficult to realize. However, the objective should be to get everybody in the organization focused on process and customers — a system where functions go seamlessly hand in hand, breaking down inefficient organizational barriers.

Team Culture, Team Spirit: Teamwork provides the necessary infrastructure for productivity gains and problem-solving within lean management and in a flattened organization structure. There are many types of teams, depending on the industry the company is in, its size, its products, its diversification, the type of manufacturing, and so on.

All teams possess certain common characteristics. The most relevant features on the pathway to success are team culture and team spirit. These two areas are too often overlooked. If they are not fostered, the team concept is severely jeopardized. The corporation's culture for team success is established by its leadership, who must convey their openness and dedication to their teams. This should be done by the following means:

Vision: Outlining the direction the company is headed over the medium to long term.

Mission: A statement of what the company is all about and how it wants to fulfill its vision.

Benchmarks: Conveying the importance of striving to be the best in cost, quality, time, and continuously improving customer satisfaction.

Ability/Capability: Making the tools available that are necessary to accomplish the set objectives.

Measures of Success/Rewards: Demonstrating the results to show that the company is within the framework of its mission and moving toward its vision. Rewarding team results.

Empowerment: Giving the teams autonomy in responsibility and decision making.

Trust: Instilling trust and building trustworthy relationships between teams and between leadership and teams.

The collection of individual experts doesn't make a team, and setting the groundwork for team culture alone does not guarantee the combined success of the team. The composition of the individual team member's attitudes and outlooks gives the team its spirit. How tasks are approached, difficulties handled, directions taken, and problems solved will determine the team's effectiveness.

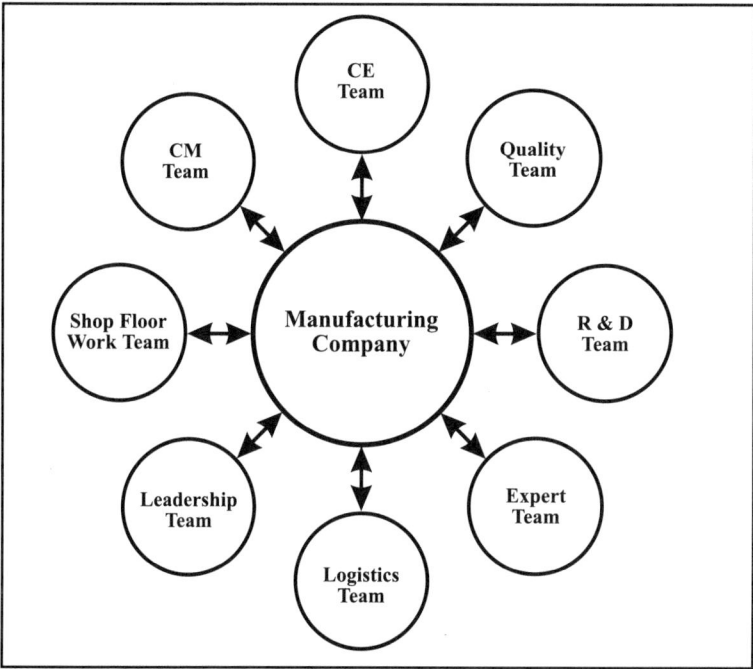

Figure 5.16 • Advanced Team Assembly

The team spirit is primarily indicated by the following.
- The leadership, team champion, and individual team members should show enthusiasm.
- There should be open, uninhibited communication within the team, with other teams, and with the company's leadership.
- The team must understand the urgency to bring about results within certain time frames.
- Developing the ability to recognize and tolerate the views of others is as important as being respectful and making compromises for the good of the team.

- There is no room for egos, jealousies, insecurities, or questionable motives and ambitions. Team members have to be goal oriented and do what is best for the company and its customers.
- Being positive and upbeat, as well as feeling that "we are in it together" and "we can do it," can lift the teams to unexpected highs.

Subscribing to the Lean and Team principles — decentralizing functions and tasks; cost, quality, and time conscientiousness; customer orientation inside and out; eliminating in-company boundaries; open, efficient communication; organizing around people and process — is the means to the company's competitiveness and survival.

Changes and Adjustments

There was a time when companies were established, an organization built up around the typical corporate functions, and a rigid structure ruled management and staff. Deviating from the "cast in stone" principles, procedures, and regulations was difficult at best because every task was approached within the traditional framework and a mindset of "playing by the rules."

Companies approaching the new millennium have to be extremely fluid and flexible, since today's business environment is dictated by fast changing market conditions. One of the most important tools for any company is to be adaptable and prepared for rapid and constant interior and exterior changes.

Revisiting "Re-engineering"

In the 1980's, General Motors realized that they were losing marketshare to other car manufacturers — primarily imports. They looked for reasons why, and they realized they had to downsize substantially across the board, flatten the organizational hierarchy, promote team concepts, be more customer oriented, shorten life cycles and the time for product development, and pursue much leaner production principles. In other words, they had to redesign their business and manufacturing processes and eliminate everything which added no value. This became known as "re-engineering."

By 1993, business had placed re-engineering high on the corporate agenda. It grew even more common in the following years to a point where it seemed to be "chic" to "re-engineer" one's business. But this tactic often led to disaster. What went wrong? What must now be done to succeed? Here are eight major factors.

1. **Poorly Conceived Planning:** Many companies go into "re-engineering" head over heels without well conceived planning and ideas. It is important to analyze the existing business thoroughly, identify what needs to be changed, and then draw up a sound plan of reorganization around the benchmarks of cost, quality, time, and continuous improvement.
2. **Patching Up Existing Ineffectiveness**: Realizing the weakness of one area within a process has nothing to do with re-engineering. In fact, this approach can make a bad situation worse for all others with relationships to this area who will have to adapt while the rest remain idle for the necessary overhaul.
3. **Too Much Emphasis On Technology**: There needs to be enough emphasis on the human aspect of the business. Before installing more or updated equipment and automation, the organization has to be in place and the team has to be functional.
4. **Wrong Approach**: Tackling the redesign of processes should not just start arbitrarily somewhere, for example, starting with downsizing and then moving on to other areas of apparent weakness. A systematic approach goes right down alongside the value-added process chain. Redesign should start at the customer's end of the chain, in order to meet their needs and expectations. The concentration must be to re-engineer for customer satisfaction. The benefits then follow along the way. Starting with the customer, work backward to best show what business process must be re-engineered.
5. **Not Enough Selling to Own Staff**: It is normal for employees to resist changes, especially something as radical as re-engineering a company. This holds particularly true when people are not told up front what is forthcoming, what is expected of them and how it will affect them. As much as possible, right from the beginning, employees should be part of designing and implementing forthcoming changes. Being aware that re-engineering their business will greatly benefit them can re-energize and revitalize the entire organization.
6. **Complexity:** Tearing down inefficient organizations and processes and replacing them must be done with simplicity in mind. Removing a managerial layer in the organization, and at the same time adding re-engineering management, defeats the purpose of

becoming leaner by downsizing. Companies that install complex information technology without redesigning and restructuring the organizational base run the risk of spending huge sums of money only to complicate everything even further.

7. **Departmental Dissolution**: Business re-engineering is about dramatic changes to achieve dramatic results. This cannot be done from department to department, or simultaneously, but rather cross-functionally and interdisciplinary, spanning all departments affected by it. It means to redesign work processes and achieve substantial improvements in performance in orders, customer service, finance, engineering, planning, production, quality, etc.

8. **Lack of Measurement**: Re-engineering involves transforming skills, technology, systems, procedures, and processes that have an impact on the culture and value of a company. These transformations have to be measured. There are a variety of measurements of re-engineering.

The measurements are to be taken during and after the re-engineering process. The former allows for corrective action before its finalization. The latter should be a comparison of the status before and after the re-engineering efforts, to see the effect the program has had. This can then be compared against other programs for effectiveness. Measuring the re-engineering process is like a barometer that indicates how and where to set priorities for improvements. It also reveals the stability of a new process in itself and compared to others.

With a firm commitment by top management and a deep involvement of the entire staff, the principles of re-engineering, if applied as outlined above, can create remarkable changes in a relatively short period of time. Reorganizing business processes offers great potentials toward competitiveness, market leadership, and customer satisfaction. The question is what comes after re-engineering when all processes have been optimized and the company is back on track. Stabilized processes have to be improved on, changes implemented, and adjustments made. The reason is that the dynamic of the market cannot be controlled. Corporations can only react. Production processes, products and marketing methods are changed; make or buy decisions are made; and in-house structures are adjusted.

Manufacturers have to make sure that their products benefit the customer, the organization is structured for the products it manufactures, and the organization is fit for the market demands. This requires fluidity and flexibility, and necessitates constant changes and adjustments.

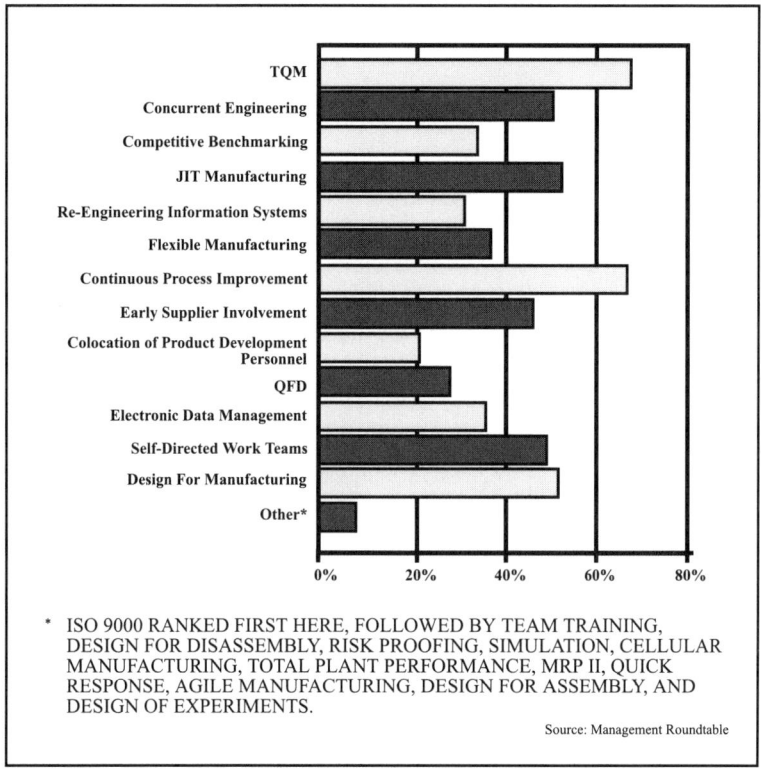

Figure 5.17 • Leading Manufacturing Principles and Strategies

The Process of Change

Change Management

Re-engineering is a massive undertaking to redirect the fortunes of corporations that adjust themselves through reorganizing. Once a different, promising course has been taken, the process of change is ongoing and never-ending because it is the foundation of sustained success. Companies have to learn and to foresee, implement, sustain, and manage change.

Successful change management is characterized by the following.

The Willingness to Change: When top management realizes that the direction of a business unit or its position in the marketplace must be altered, it must demonstrate a willingness and commitment to change, plus it must show a direct involvement to rally its staff. Corporate management must be prepared for any sudden shift, and establish the in-company culture which is reflected by the immediate acceptance of the change throughout.

The Strategy of Continuous Improvement: A business that institutes an ongoing quest for doing things better implements systems and processes in order to often forego dramatic changes, since they could be well on their way by the time redirections are needed.

The Corporation's Competence: The degree of the abilities and capabilities of a company reflects how top management promotes continued education, the free flow of ideas, and the acquisition of updated machinery and equipment. Management must know its staff's strengths and areas of expertise and make the best use of them.

The Importance of Being Proactive: Being caught "off-guard" by market events puts a company into a reactionary position. It is forced to play "catch-up." That can be an uphill battle and could strain the organization just to close the gap with its competitors. Once a company falls behind it makes it extremely difficult to take or regain a leadership position. Proactive changes are much easier to implement, and are golden opportunities to leapfrog the competition. Enablers for proactive change are top management's vision of the future; as well as key personnel's deep involvement in the day-to-day happenings of the market, their foresight of what others could be doing, and determining their standing relative to the competition. Looking for best practices and performing benchmarking are the cornerstone for proactive change. At the same time, continuous improvements can be made on the relevant processes. Interdisciplinary teams can act as think tanks, and brainstorm independently for new pathways and options. There must always be a clear road map for innovative approaches to lead to a proactive change matrix.

The Systematic Approach to Change

The process of change must be pragmatic and systematic. It basically involves the following questions.

- What has to be changed? It could be the place of business, the product, a process, communications, the organization, etc.
- Why does it have to be changed? A loss in marketshare, outdated technology, poor response time, and in-company logistics not being up to par.
- How should it be changed? Through concurrent engineering, benchmarking, project management, and market studies.
- Who has ownership of the change? Top management, in-company working teams, multidisciplinary teams, outside consultants, or integrated suppliers.

The progress of the interdependent change cycle — its cause, objective, strategy, and action (See Figure 5.18) — must be monitored and updated for all team members and top management, in order to set priorities and decide early on (and at any stage of the process) what direction to pursue more vigorously and when to "throttle" back. This prevents the change process from moving into a dead end. It also makes the best use of the two resources: people and financing.

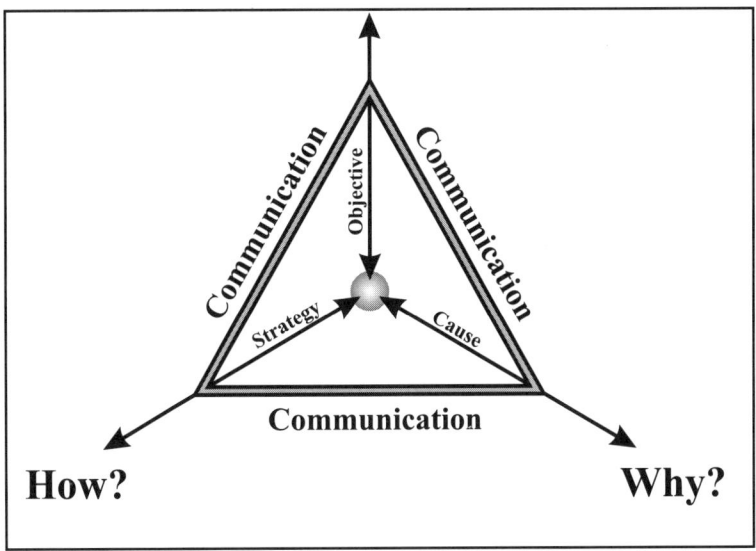

Figure 5.18 • The Diamond of Change

The Most Relevant Change Strategies

Besides full-blown business re-engineering and the process of major changes such as logistics, manufacturing, administration, etc., the most relevant strategies for change embrace innovation and attitude.

Innovation

Clearly, any manufacturing company at some point has its products face a saturated market, customers demanding different products, or its competitors dividing the market among themselves, thus forcing it to abandon the marketplace. The surest way to be a perpetual contender, and secure marketshare or possibly market dominance, is innovation. Management has to pursue the culture of permanent change for the never-ending quest for product, process, and marketing innovations. Innovation driven companies are creative in strategically competing in the marketplace. They can change from focusing on an original new idea to servicing the product in the field, and every step in-between.

The change process starts with teams that question every aspect of the "status quo," network all causal interrelations, and connect them dynamically to anticipatory developments. Once a new, realistic idea has been created, it is passed on to the research and development team. Here it becomes necessary to change existing methods and research processes to develop a prototype product, thus setting off the value-added chain. Often the development of a new or improved product identifies a weakness in the current manufacturing process, so that the necessary changes can then be implemented before the final product design. Changes in one area can have the snowball effect, leading to more changes in other areas of existing processes. This is a decisive factor of interdisciplinary team-building.

Once the changes for innovative product development and manufacturing processes have been finalized, the conversion to a marketable product is the next step. Innovative products and processes have value to the manufacturer only if they can be successfully marketed; and when the new, innovative product hits the market, more ideas and changes have to be contemplated to secure broader acceptance of the product, find related applications, and possibly connect to a different customer base.

The manufacturer of a newly introduced product has to combine all its resources in order to avoid being pushed over by its competitors. Changes for product Mark II replacing Mark I have to be worked on with the inception of the latter. To fend off "creative copycats," the manufacturing company might be willing to grant Mark I licenses to competitive or semicompetitive marketers to accelerate the product's market acceptance, to keep competition at bay, and to stay in control of the market. This approach necessitates dramatic changes in marketing, particularly affecting distribution, advertising, and product support.

Another proactive change within the manufacturing company is the further development of an original innovation by another company, which is essentially an advanced version of the original product. This "creative copying" involves even more time, and disciplines manufacturing companies to be patient for external changes while the relevant internal planning for product development and machining manufacturing processes is well under way. When top management sees the window of opportunity for their advanced product, it is ready for potential external (marketing) changes. This occurs at a time when the market is beginning to accept the new product, however, the innovative newcomer might better understand what the customer really expects the product to be.

Change is the underlying connotation of innovation. It usually affects all other changes automatically or by necessity and opens up new avenues and opportunities for all corporate areas. Promoting creativity and innovative ideas means being prepared for immediate and constant changes affecting the entire enterprise.

Attitude

If innovation is the incarnation of change, then organizational mindset and attitude are the prerequisites for it. The most precious resource of any company is its people. However, while not every staff is excellent, a team is resilient to any unforeseen change and can be integral to implementing well-conceived change. Developing, multirating, and leveraging a company's intellectual assets is the key to the success of any necessary change. Only through sustained efforts on the part of management and employees can an organization be fit for change. Coping with change, learning to accept it, and viewing it as a the chance for future growth and success, as well as an opportunity for personal betterment and security, is achieved by broadly disseminating the change mission and by minimizing possible disruptions from the existing tasks. The pathway to optimal performance of change management has three stages: preparedness, willingness, and ability.

Preparedness

It is the responsibility of management to introduce change and inform the workforce of how it will affect them so that they can be prepared for it. Any change is generally regarded by the individuals and teams as a disruption of known daily routines. When a change is to be made, management has to be sure that their staff is ready for the change to be implemented. There must be a constant flow of information throughout, and everyone in the organization must understand the need for the change.

Management also needs to share the magnitude of the change, and give an estimate of the time it will take to complete the task. A ready-for-change organization will proceed with confidence and accomplish the change more quickly. A sure and swift change procedure helps to make workplace change easier. A company's degree of preparedness for change is directly related to management's communicative, open, and understanding leadership.

Willingness

Being willing to change requires overcoming the natural resistance to change and the desire to maintain the status quo. Navigating the difficult course of necessary and constant changes, with employees willing to embrace them, can be accomplished in the following ways.

- Make people feel safe during and after the process (job security).
- Communicate the advantages (growth, marketshare, etc.).
- "Being in it together" reassures that everybody is affected by it.
- Secure adequate resources and staffing.
- Stress teamwork, but recognize the individual's value.
- Practice the 3-R concept (responsibility, recognition, reward).
- Encourage, support, and coach — before and during the process.
- Develop strategies to make changes desirable.
- Be open for suggestions that might redirect the process in midstream.
- Provide constant feedback to the teams and individuals, so that they know how they are doing during and after the change.

Ability

The performance of the people involved in the change process depends to a great extent on their ability. Ability is the sum of skills and talents, and the power to perform certain tasks. Knowledge and expertise have to be acquired, developed, and enhanced. The skills and power for maximum performance in the change process derive from a "learning" company — one that emphasizes the following:

Learning: Institute seminars and training classes for continuous education. In a learning organization, people continually discover new possibilities and opportunities. Cross-training at different assignments increases depth and aptitude, and broadcasts knowledge throughout the organization.

Empowerment: Enable people to control their assignments. As teams, they manage themselves and are authorized to make decisions. Independence should be allowed to a high degree, but it must be coached by management.

Fitting: Teams can only work and perform satisfactorily if they work as a harmonious "one." Management wants as much knowledge and expertise in the working group as possible; however, it has to make sure that the individual personalities combine well to form group strength. It is equally important to provide feedback to the teams so that they know how their efforts fit in the overall scheme of the change process. Bringing everybody up to date on the progress made reinforces the group's efforts.

The more knowledge a company disseminates amongst their employees, the more able the organization becomes. Accelerated learning leads to increased efficiency of ever more frequent change processes.

Adjustments

The course of change must often be redirected due to unexpected obstacles, different ideas and approaches, or new objectives; and adjustments have to be made in disciplined, organized, and clearly defined ways. Stumbling blocks along the way must be carefully analyzed to decide which adjustments to make. Sometimes, obstacles are self-inflicted. That is, teams perceive the change to be too complex, and so half-heartedly meet the challenge, or they rush in too fast because they are eager to reap the benefits and hastily implement the change without proper planning.

Whatever the reason for a sudden roadblock, the change team will have to regroup to find responses to the questions raised: What stage are we at now? Where do we want to get? What is a better route to take? At this point, a renewed, strong commitment by the leadership of the company is required to encourage new ideas and alternative actions. The key is to maintain morale and to foster creativity.

The directive for change must be given before a real crisis occurs. Adjustments also must be made before it is too late, or before a complete overhaul becomes too costly and a huge liability for the organization. A crisis situation can easily be avoided by using foresight during the planning stage. Anticipating the worst helps in mapping out plans of prepared alternative actions. It is easier to prevent fires than to extinguish them.

The company's leadership must be involved in the change process in order to prevent small problems from turning into a full blown crisis. Adjustments will then have to be made after completing the change process and implementing the new course of action because the party for which the change was made may not be completely satisfied. If the change affects units *within* the company — for example, the machining area of production or design processes of product development — adjustments can more easily be made because of the internal relationship. But if the change affects the company's external activities — particularly its customers — the problem is compounded by the fact that, on average, only 30% of customers voice their complaints. Once the manufacturer adjusts a product feature, it is regarded by the majority as long overdue, although the company reacted immediately when the complaint became known. This time lag can have a devastating effect because some of the customers might already have acquired a more suitable product from a competitor. Now the manufacturer is forced to control the damage. Employing the following three basic rules can be very helpful.

1. Thoroughly analyze the situation. Try to get to the core of what went wrong, why, and at which stage of the change process. Ask customers for their input and let them know you are very sincere in changing the situation.
2. Correct the problem with speed. Decisive actions and speedy adjustments are impressive countermeasures and can win over dissatisfied customers with ease.
3. Use the error or fault as an opportunity for improving related processes or strategies. Find out what went wrong, and how to prevent it from recurring. Other weakness of the organization might become apparent. Proactively making adjustments can thus prevent other potential, possibly even more detrimental, problems.

With regard to customer-oriented, proactive adjustments, constantly measuring the degree of customer satisfaction is a tool that keeps the finger on the pulse of the marketplace and offers opportunities for corrective measures right at the outset. The objective is not to groom a department for "complaint management" (a reactive approach), but rather to develop a dedicated system that compiles and analyzes data and feedback, identifies potential problem areas, and recommends solutions.

AGILITY – THE JOURNEY OF CONTINUED COMPETITIVENESS

Drivers of Change and Competitiveness

In the 1980's the U.S. government realized that there was a major shift underway toward a more global economy and, therefore, a much more competitive manufacturing environment. Searching for ways to enhance the nation's competitiveness through new manufacturing strategies, the Secretary of Defense asked Lehigh University's Iacocca Institute to work on the project. The resultant book titled, *Twenty-First Century Manufacturing Enterprise Strategy*,[5] summarized the brainstorming of teams of industry executives. They used the term "agile" to describe the key elements of a manufacturing company's competitiveness. By studying and benchmarking Japanese auto manufacturers, the findings concluded that when applying the recommended manufacturing principles, U.S. manufacturers could even leapfrog Japanese manufacturing success.

Concurrent engineering is the first step to being agile — by relying on process-oriented cross-functional teams, managing product design and product development, tearing down departmental walls and cutting across the vertical functions, and going beyond the boundaries of their own corporation to include suppliers and strategic partners. When concurrent engineering was turning out costly revisions due to afterthought on the production floor, the next step was to interlink the participants of teams through advanced information systems. "Doing things right the first time" was the harbinger of things to come.

• • • • • •

[5] Goldman, Steve, Preiss, eds., *Twenty-First Century Manufacturing Enterprise Strategy*, 2 vol., Lehigh University Press, Lehigh, PA., 1991.

The manufacturing forum now defines agility as the ability of a company to adapt and thrive in a business environment of continuous and unpredictable change. To achieve this, traditional methods of production of consumer products had to be revised. The change to mass customization was the next step on the road to agile manufacturing.

In the wake of much needed cost cutting, manufacturers started to institute Lean Management and Lean Production, in which just-in-time deliveries to manufacturing plants have become standard practice. Demands for faster response time and shorter lead times will increase, so flattening hierarchies and working in teams is required to create structural flexibility and to be prepared for unanticipated (but necessary) changes. By decentralizing decision making and the flow of information, organizations can process more and faster.

Rapidly changing technology adds pressure on manufacturers. More emphasis is put on research and development in the quest for new, innovative products. Then the key is to have the ability to bring the product to market quickly and possibly create offshoots for other market niches based on practically the same innovation. A growing number of enterprises are spinning off products that they feel can be better made by suppliers; and while their own content of the finished product shrinks, the role of subcontractors gains in importance. The formation of partnerships is a direct result of the manufacturers' strategy to concentrate on their core competency and to produce products they consider themselves to make best.

As OEM's commit themselves to producing their products in other countries, to be close to the market they serve, and to have an economic advantage, some of their suppliers go with them and continue their close relationship offshore. When manufacturing companies go global, their organizational and manufacturing activities have to be coordinated. Technology transfer, proven processes, continuous improvement, level of knowledge, uniform quality standards, etc., have to be managed. Agility is basically for change and competitiveness. It does *not* imply injecting all current principles into an existing enterprise and then expecting these changes to secure instant success and lasting competitiveness. The principles and strategies of agility are fluid, and are themselves subject to changes and adjustments.

Key Attributes of the Agile Manufacturing Company

The process of agility encompasses many areas:
- concurrent engineering/manufacturing,
- total quality management,
- lean management/production,
- (mass) customization,
- integrated information systems,
- team and process orientation/change management,
- customer satisfaction,
- continuous improvement,
- learning organization,
- flexible and versatile processes, and
- core competency which is knowledge and technology driven.

Characteristics Of Agility

Agility within manufacturing means being able to adapt and to work within mass production or a purely customized manufacturing environment. Its versatility is such that its characteristics, over time, can be implemented without disturbing the current state of the manufacturing enterprise. One area of change invariably stimulates others to eventually create an agile manufacturing company, whose attributes make it resistant to the vagaries of the market it competes in.

Developing Competency

The core competency of any company is its knowledge and its talent. The degree of knowledge is the result of the strategies set forth by top management, the determination to embrace learning across the organization, the ability of the company's employees, and the ongoing search for more skills and expertise.

The essence of agility is adaptability to change. Hence, part of the strategy of top management has to be to promote responsiveness to changes, and to ensure that the organization adapts to changes rapidly, with precision, and with tightly held ideas to create a responsive team as well as company-wide ownership of ideas. Changes can occur unexpectedly, or they can be anticipated and even induced. The low frequency of proactive and reactive changes can be an indication of a "fragile" company, while a company with a low proactive rate combined with a high reactive rate can be called "opportunistic." Innovative companies usually have a high rate of proactive changes, however, they are low on reactive changes. An agile company, on the other hand, is characterized by a higher frequency of both proactive and reactive changes. No matter what change has to be dealt with, the organization must be able to adjust to, correct, and enhance the new situation. A learning company is one that encourages, injects, develops, and disseminates knowledge throughout the entire organization. Such a company is cognizant of the fact that knowledge itself is subject to change and must, therefore, be renewed, refreshed, refocused, and reinterpreted.

Technologies, markets, applications, and processes change rapidly, and they affect a company's competency. An agile company is always in a position to abandon certain techniques, products, or systems in favor of more appropriate ones. Shifts in markets, products, or processes enhance its knowledge and broaden the base of its competency. A learning company prepares its staff proactively for unforeseen changes through dynamic workshops that play out complex scenarios in "think tanks" to obtain new wisdom or ideas. Once new knowledge is discovered, it must be shared within the organization. The mix of established and new knowledge makes it a learning company. A company that has the ability to convert such knowledge into practical use — and apply it — is competent. Staying competent is a matter of timely adjustments to rapidly changing technology. Finally, a competent manufacturing company will know the fastest, best, and least expensive way to make a product, even if it means adopting technology from the outside. Such resourcefulness can be an invaluable asset for a competent, agile manufacturing company.

Customer Information/Communication

The importance of communication within any company has already been discussed. The agile enterprise, by nature, interfaces more directly with the outside. In fact, it is primarily the erratic market shifts and frequent changes in customer tastes and expectations that induce such a company to embrace the principles of agility. This puts extraordinary value on external communication. The two areas that are of special relevance are "customization" and "closeness" to the customer; and effective communication with the customer is the prerequisite for successful customization.

Service automation is a first step to mass customization (see Figure 5.19). A company limits itself to low variety, standardized products that

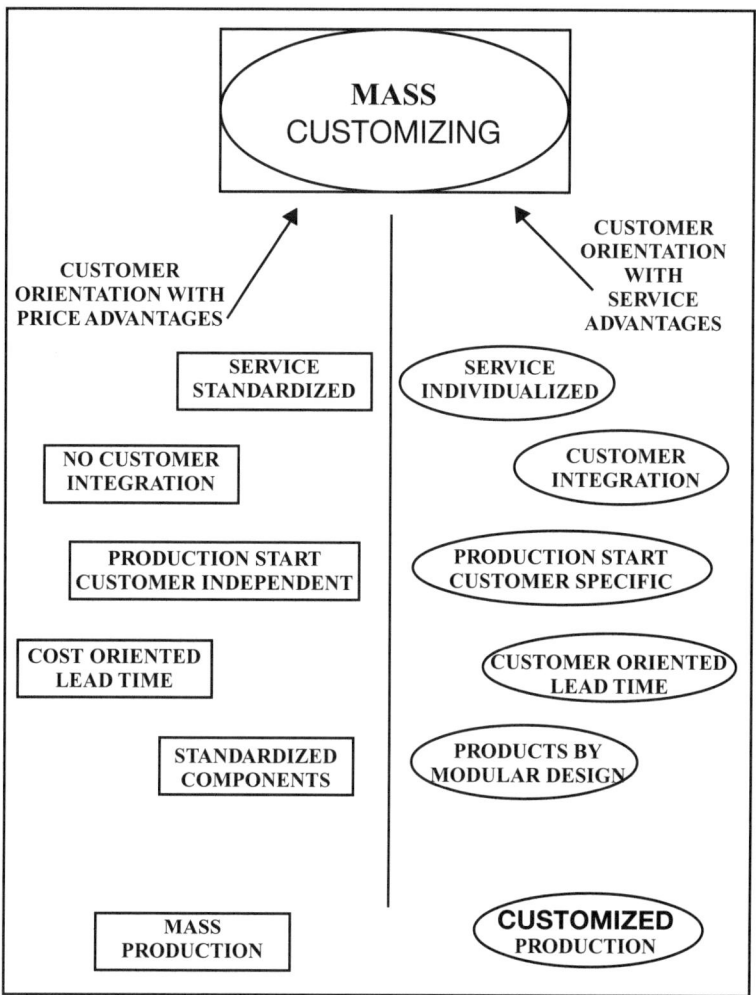

Figure 5.19 • Mass Customization

its customers prefer. Constant dialogues, valuable feedback, and shared values with its customers invites the manufacturer to supplement the supply of parts with intense service such as technical consultation, demo rooms training, and warranty agreements. Attractive, overall packages are devised and implemented by producer and customer. Any adjustments to current agreements are made with ease, at favorable cost, and quickly through an effective two-way communication system.

"Self-customizing" takes the production of products based on constant customer information and communication even further. Here, the mass customizer produces mass products with built-in flexibility. With this concept, the customer literally tells the manufacturer what "built-in" flexibility is expected. Through a constant dialogue, the customer enables the producer to know what to customize. A typical example is the computer industry, which offers standardized computer hardware with optional software packages. The importance of a close relationship with the customer cannot be stressed enough. The information flow has to be fluid, and the communication channels open. The more complex the product and the more competitive manufacturers vie for their marketshare, the more important it is to pass on the information customers really want and really need. Only then can the producer establish a relationship for correct and necessary feedback. A systematic and continued analysis of the rapidly changing market parameters and the resulting customer preference can only be obtained through very close customer relationships.

Every manufacturer, over time, is faced with the following, real life situations.

- A product is rendered obsolete and reaches the end of its life cycle.
- Product demand is much higher than expected.
- Product demand is lower than the optimistic forecast.
- Constant product changes interrupt regular production flow.
- Frequent process changes occur during high peak production runs.
- Small and large part runs fluctuate.
- Varying standardized and customized part lots must be accommodated.

Manufacturing with agility means that equipment and processes are in place, and are capable of adapting to the following variations in manufacturing without sacrificing product quality or manufacturing productivity:

- Machines in reconfigurable factory layouts.
- Movable machining centers forming highly flexible cells.
- Machine tools and cutting tools made with modular components.
- Modular, versatile part fixturing.
- Machining processes with high repeatability accuracy.
- Uniform machine control systems (open architecture).
- Trammable precision cutting tools.
- Multiskilled workforce in all operations.

- Process simplicity.

(*Note*: Adaptable manufacturing systems do not necessarily have to be fully automated. Flexibility and adaptability are provided by a skilled, motivated workforce, and by adjusting processes as needed to meet different production requirements.)

Measuring the Progress of Agility

Knowing where one stands in striving for agility is important to take appropriate actions in areas that need more attention. The following are indicators of corporate-wide progress.
- The time it takes to produce a new customized product.
- The percentage of customized products compared to standardized products.
- The percentage of one's own content of a finished product.
- The total number of new customers and their feedback of service capabilities.
- The number of new products introduced ever since the inception of agility.
- The number and types of niche markets.
- The status on acquisitions, mergers, and partnerships.

Agile manufacturers continually measure progress against market benchmarks.

New Frontiers — New Challenges

Agile companies are never static — they are flexible within themselves and in relation to the outside. They observe and analyze, devise new plans and strategies, and convert them into action. Above all, they are not reactive. They make it a point of being involved and ahead of situations in areas such as changing customer needs, environmental and health issues, new innovative or evolutionary product and process technology, technology substitutions, etc. See Figure 5.20.

Today's manufacturers are getting ready for the "virtual" company. Virtual companies are enterprises that are assembled on an interim basis (or temporarily) to produce products or provide services only long enough to satisfy the original need and complete the mission they set for themselves.

Each of the companies contributes people or equipment that the other does not have and has no intention of acquiring. Foregoing formal, joint ventures or mutual acquisitions, they keep their financial and corporate

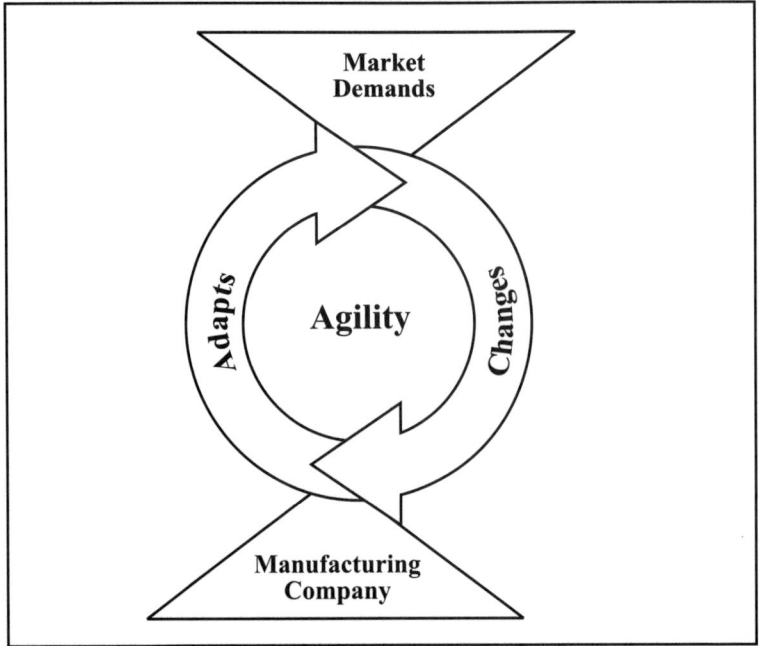

Figure 5.20 • Agility, Step By Step

independence. It is an interesting concept in cases of customized products with a predictable lifetime. Such a partnership can even be forged between otherwise competitive companies. While they compete against each other in their traditional market with competitive products, they might form a "virtual" company to capture a new, additional, albeit temporary market. Rapid technological changes and the availability of comprehensive expertise and knowledge in too many fields makes the "virtual" enterprise a viable proposition.

Cooperative value-adding is an interesting offshoot of agility, and a promising strategy as we enter an era of new dimensions in manufacturing. Roger Nagel, Director of the Iacoca Institute, says, "Agility is not something you buy into or don't buy into. It is coming because of market forces. Agile manufacturing will be more than a competitive advantage in the future, it will be a condition of survival."

CONCLUSION

Human creativity will never cease. Inventions by great individual minds will always provide the impetus for progress. Often, though, systems, products, processes, and techniques are so complex that innovation is only possible through the efforts of a working group in a setting of dedicated teamwork.

It is prudent to be flexible and open minded in the search for more productive and economical solutions. The mere knowledge of the availability of new technology is meaningless unless it is converted into practical use. Surprisingly, updating organizational abilities and capabilities can often be done through minor and piecemeal adjustments, rather than complete overhauls in short time spans.

It is apparent that the implementation of advanced strategies, techniques, and processes not only impacts the areas they were intended for, but also other directly and non-directly related sectors.

Progressive manufacturing companies inspire the communication of information that sets the stage for other new ideas and developments. Certainly, there will always be companies that bow out of production or the market entirely, and it might very well be that the manufacturing arena will shrink in terms of its participants. Manufacturers are faced with great challenges in the next few years, but they will also be exposed to enormous opportunities. Applying systems processes and techniques to eliminate current waste in all facets of business is an untapped resource and promises the creation of a better world that will benefit everyone.

Therein lie the merits of the new dimensions in manufacturing.

References and Suggested Reading

Akao, Yoji (Ed), **Quality Function Deployment,** Integrating Customer Requirements into Product Design, Productivity Press, 1988.

Bakerjian, Mfge.E, Ramon (Ed), **Tool and Manufacturing Engineer's Handbook,** Volume 7, Continuous Improvement, Society of Manufacturing Engineers, 1993.

Burgelman, Robert A., Maidique, Modesto A., **Strategic Management of Technology and Innovation,** Irwin, 1988.

Camp, Robert C., **Benchmarking,** The Search for Industry Best Practices that Lead to Superior Performance, Quality Press.

Carr, David K., Hard, Kelvin J., Trahant, William J., **Managing Change Process,** A Field Book for Change Agents, Consultants, Team Leaders and Reengineering Managers, Coopers and Lybrand/Center of Excellence for Change Management, 1996.

Hronec, Steven M., **Vital Signs,** Using Quality, Time, and Cost Performance Measurements to Chart your Company's Future, Arthur Anderson & Co., 1993.

Juran, J. M., Gryna, Jr., Frank M., **Quality Planning and Analysis,** From Product Development Through Use, McGraw-Hill, 1980.

Lundy, James L., **TEAMS/Together Each Achieves More Success,** How to Develop Peak Performance Teams for World-Class Results, The Dartnell Corp., 1994.

Oleson, John D. (1998), **Pathways to Agility,** Mass Customization in Action, John Wiley & Sons, 1998.

References

Peters, Tom, **Thriving on Chaos,** Handbook for a Management Revolution, Harper & Row, 1987.

Porter, Michael E. (Ed), **Competition in Global Industries,** Harvard Business School Press, 1986.

Smith, Graham T., **Advanced Machining,** The Handbook of Cutting Technology, IFS Publications/Springer Verlag, 1947.

Sushkin, Marshall, Kiser, Kenneth J., **Total Quality Management,** Ducochon Press, 1992.

Tompkins, Ph.D., James A., **Winning Manufacturing,** The How-To Book of Successful Manufacturing, Tompkins Associates, Inc., 1989.

Weck, Manfred, Eversheim, Walter, Koenig Wilfried, Pfeifer, Tilo, **Production Engineering,** The Competitive Edge, Butterworth Heinemann, 1991.

INDEX

A

ability, 168
adaptive control, 87
advanced cutting geometry, 32
advanced planning
 and scheduling (APS), 8
advanced supply train, 139
agile manufacturing, 148
agility characteristics, 173
aluminum matrix composite, 25
aluminum oxide
 (Al_2O_3), 23, 26, 30, 47, 79, 92
aluminum, 24, 51
applied tool management, 34
attitude, 167
automation, 146

B

balancing tooling systems, 35, 37
bearings, angular contact, 3
bearings, hybrid, 3
bearings, hydrostatic/hydrodynamic, 3
benchmarking, 153, 158
benchmarks for manufacturing, 104
benchmarks for performance, 117
borazon, 63
bore quality, 60

C

carbide cutting tools, 78
carbide insert coatings, 50
ceramic cutting tools, 79
cermets, 30, 47, 79
change diamond, 165
change management, 163
chip control, 52, 82
chip control in
 high speed machining, 7
chip volume, 2
choosing a supplier, 131, 132
circular interpolation, 95
circular milling, 95
coated carbide cutting tools, 79
complete customer orientation, 156
composites, 25
concurrent engineering trilogy, 105
concurrent engineering, 105, 153
concurrent manufacturing, 110, 114
concurrent manufacturing
 benefits, 127
continuous
 improvement, 109, 113, 156, 164
coolant cost in machining, 44
coolant functions in machining, 43
cost to design (CTD), 104
cross-functional teams, 111
cryogenic cooling system, 55
cubic boron nitride (CBN), 80
cutting chips in dry machining, 52
cutting forces, 2
cutting material hardness, 47
cutting material tensile strength, 48
cutting tool life, 2
CVD coating, 25

D

data management, 121
design for
 manufacturing (DFM), 104, 143
design to cost, 143
diamond like carbide (DLC), 49
dimensional tolerancing, 78
drilling cast iron and steel
 at high speeds, 22
drilling composite materials, 62
dry and high machining, 63
dry drilling, 59
dry fineboring, 56, 58
dry machining, 46
dry milling, 62
dry reaming, 56
dust in dry machining, 52
dust in high speed machining, 7

E

enterprise resource planning (ERP), 8

F

face milling cast iron and steel, 21
face milling cutting speeds, 39
failure mode and effect analysis (FMEA), 122, 153
feed and speed ranges, 20
filtration systems, 52
fineboring, 97
fixturing in high speed machining, 8
flat hierarchy, 158
flexible manufacturing, 85, 148
flexible manufacturing systems (FMS), 146
form clamping, 11

G

Garvin, David, 124
global manufacturing, 122

H

hard turning, 53, 91, 92
heat in dry machining, 52
heat in high speed machining, 7
heavy metal, 82
high cutting speed, 2, 49, 82
high cutting speeds in dry machining, 49
high depths of cut in dry machining, 49
high feed rates in dry machining, 49
high speed machining, defined, 1
horizontal corporation, 157
HSK clamping system, 9, 10
HSK maximum speed, 17, 19
HSK tooling interface, 8
HSS, 47
HSS PVD coated tools, 48

I

Iacocca Institute, 171, 178
inconel 718, 23
in-house suppliers, 133
innovation, 166
ISO 9000, 131, 135
ISO-taper, 8

J

just-in-time delivery (JIT), 112, 133, 156

K

Kitano, Mikio, 34

L

laser assisted turning, 54
laser supported cutting, 53
lean and team principle, 160
lean concept, 155
lean manufacturing, 148
linear scales, 85
lost foam casting, 111

M

magnesium, 24
magnesium, critical temperature, 25
manufacturing cells, 8
mass customization, 174
minimum volume lubrication (MVL), 63
modular sourcing, 137
molybdenum disulfide (MoS_2), 48
motor, linear, 4, 5
motor, servo, 4, 5
multipurpose tooling, 35

N

near-dry drilling, 67
near-dry fineboring, 70
near-dry machining, 63, 65
near-dry precision fineboring, 71
near-dry reaming, 68, 69
nickel, 34

O

O.D. reaming, 91
oil/air mix for cooling, 66
one pass machining, 76, 82
open architecture, 5, 6, 147
optimizing processes, 116
optimum cutting speed, 41
OTT-System, 18
outsourcing, 118

Index • 185

P

pallet exchange, rapid, 7
PCD tipped drilling, 62
physical vapor deposition (PVD), 47
plug-and-play principle, 146
polycrystalline cubic boron nitride (PCBN), 23, 26, 29, 53, 92
polycrystalline diamond (PCD), 23, 26, 27, 80
power clamping, 11
preferred supplier, 134
preferred *systems* supplier, 134
process monitoring, 85
product life cycle, 107
pull concept (inventory), 140

Q

quality function deployment (QFD), 143, 144, 153

R

reaction bonded silicone nitride (RBSN), 55
reaming cast iron and steel at high speeds, 22
re-engineering, 160

S

S shaped product maturity curve, 123
service automation, 174
silicon-nitride (Si_3N_4), 23, 26, 30, 47, 53, 79, 92
single sourcing, 137
SPC/CPK based manufacturing, 119
superalloys, 22
supplier demands, 134
supplier integration, 141, 156
supplier involvement, 130
supplier opportunities, 135
supplier types, 131, 132
surface finish, 90
surface quality, 2
systems suppliers, 132, 139

T

Taylor curve, 40
team concept, 106
technology triumvirate, 3
temperature distribution in cutting, 46
thermostress, 52
thread milling, 96
3-r concept, 168
tier 1 supplier, 131
titanium aluminum nitride (TiAlN), 47, 48,63
titanium aluminum vanadium (Ti-6Al-4V), 55
titanium carbonitride (TiCN), 47,63
tool change, fast, 7
tool deflection, 33, 83
tool failure, 37
tool overhang, 33
torque in high speed machining, 11, 12
total indicator readout (T.I.R.), 11, 14
total product tolerance, 148
touch probe, 87
touch sensor, 87
tungsten, 34
turning cast iron and steel at high speeds, 21
turning using dry machining, 53
turnkey supplier, 133

U

unbalance in high speed machining, 15, 16
University of Nebraska, 55

V

value engineering, 153
vibration in machining, 82
virtual companies, 177

W

wet machining, 43
window supplier, 112
world class manufacturing, 126, 128

Z

zero defect machining, 78
zero defect, 156